THE DNA FIELD
and the
LAW OF RESONANCE

THE DNA FIELD
and the
LAW OF RESONANCE

CREATING REALITY THROUGH CONSCIOUS THOUGHT

PIERRE FRANCKH

Translated by Aida Sefic Williams

Destiny Books
Rochester, Vermont • Toronto, Canada

Destiny Books
One Park Street
Rochester, Vermont 05767
www.DestinyBooks.com

Text stock is SFI certified

Destiny Books is a division of Inner Traditions International

Copyright © 2009 by KOHA-Verlag GmbH Burgrain
Translation copyright © 2014 by Inner Traditions International

Originally published in German under the title *Das Gesetz der Resonanz* by KOHA-Verlag
First U.S. edition published in 2014 by Destiny Books

All rights reserved. No part of this book may be reproduced or utilized in any form or by any means, electronic or mechanical, including photocopying, recording, or by any information storage and retrieval system, without permission in writing from the publisher.

Note to the Reader: This book is intended as an informational guide. The remedies, approaches, and techniques described herein are meant to supplement, and not to be a substitute for, professional medical care or treatment. Tey should not be used to treat a serious ailment without prior consultation with a qualified health care professional.

Library of Congress Cataloging-in-Publication Data
Franckh, Pierre.
  [Gesetz der Resonanz. English]
  The DNA field and the law of resonance : creating reality through conscious thought / Pierre Franckh.
      pages cm
  "Originally published in German under the title Das Gesetz der Resonanz by KOHA-Verlag."
  Summary: ""A practical guide to unlocking the powers of our DNA to manifest health, wealth, and happiness" — Provided by publisher" — Provided by publisher.
  Includes bibliographical references and index.
  ISBN 978-1-62055-347-3 (paperback) — ISBN 978-1-62055-348-0 (e-book)
  1. New Thought. 2. Mind and body. 3. DNA. 4. Success. I. Title.
  BF639.F68513 2014
  131—dc23
                                                                                           2014018571

Printed and bound in the United States by Lake Book Manufacturing, Inc.
The text stock is SFI certified. The Sustainable Forestry Initiative® program promotes sustainable forest management.

10 9 8 7 6 5 4 3 2 1

Text design by Virginia Scott Bowman and layout by Debbie Glogover
This book was typeset in Garamond Premier Pro with Tussilago and Helvetica Neue used as display typefaces

To send correspondence to the author of this book, mail a first-class letter to the author c/o Inner Traditions • Bear & Company, One Park Street, Rochester, VT 05767, and we will forward the communication, or contact the author directly at **www.pierrefranckh.com**.

Contents

Introduction: Who Would You Like to Be?   1

PART 1
Discoveries That Changed Our World

Each Era Has Its Own Breakthroughs	7
What Is Resonance, Really?	8
How Do We Really Send Out Our Wishes?	10
How Can Our Beliefs Change the Outside World?	17
Can We Impact Our Cells through the Power of Thought?	23
Can We Impact Far Distant DNA through the Power of Thought?	30
How Do Our Wishes Become Reality?	35
How Do Affirmations Work?	41
Can We Create a New Future through the Power of Thought?	45
Can Matter Also Be Changed by the Power of Thought?	52
Connected to Everything	56

PART 2
Effective Ways to Come into Resonance with Your Wishes

What Resonance Field Do You Exist In?	64
When Resonance Fields Hinder Your Further Development	68
The Gift of Mirror Neurons	75
Surrender to the Field of Resonance	82
How Does Your Resonance Field React to News Media?	87
The Brain Is Malleable	93
The Power of Affirmations	98
Build an Image of Your Wishes	103
Accelerate the Construction of Resonance Fields	108
Using the Healing Power of Sound	112
Give Yourself the Gift of Recognition	116
Can One Heal Oneself with the Power of Thought?	122
The Power of Attraction between Two Soul Mates	128
How Do You Build a Resonance Field for Your Ideal Partner?	131
Coming into Resonance with Yourself and Your True Feelings	140
Using the Power of Prejudgment in a Positive Way	143
Can One Bring About Enduring Change in the World through Positive Thinking?	148

How Can Old Beliefs Be Transformed?	153
Every Forgiveness Brings a New Beginning	159
Change Your Past	164
Do Not Overstrain Yourself	168
Afterword: Simply, Thank You	173
Resources	174
Index	177

*The exceptional power
That lies in the law of resonance
Is probably one of the biggest discoveries of our lives.*

*When wishes have not yet been fulfilled,
When yearnings have been unfulfilled,
When things occur in our lives
That we certainly do not want,
When we are unlucky
Or when we must admit defeat,
We find the key to this in the law of resonance.*

*When we begin to fathom
How to use the law of resonance,
Everything becomes possible.*

INTRODUCTION

Who Would You Like to Be?

In life, there are no solutions. There are forces in motion: one must create them, and the solutions will follow.

ANTOINE DE SAINT-EXUPÉRY

If you could be who you always wanted to be, who would you be? If there were no boundaries, no reservations, no one saying your deepest desires and wishes overreach or are entirely mad, laughable, or presumptuous, who would you be? If you could truly be the person you always wanted to be, with your path unobstructed and all doors wide open, who would you be?

These kinds of questions kept me busy even as a young boy. What we wanted to be, even then, at such a young age, had tremendous significance for me and my friends. Without really knowing it, my friends and I were already in the age range where we were setting a course for our futures.

Throughout my life, the importance of this question never diminished for me, and today I ask it of myself more frequently than ever before because the respective answer defines my life. My response informs my yearnings, my decisions, my judgments, my convictions, and my development. And as long as my surroundings change accordingly, ultimately my entire life is changed.

Yet when I ask other adults to answer this same question, I am often confronted with misunderstanding and skepticism. Many adults find the idea of just *thinking* about this silly; they no longer occupy themselves with this question. Why is this so? Is it because when you are utterly convinced that you cannot change your life, it is arduous just to think about asking yourself "Who would I like to be?"

It is a tragedy when one is not in a position to change or alter the circumstances of one's life simply because of some rigidly held beliefs, because it is a fact that our beliefs write the script of our lives.

The most recent discoveries of quantum physics, quantum biology, modern mathematics, and epigenetics (the study of mechanisms that regulate gene activity) clearly reveal that the power of our belief systems creates the person we believe we are—from our health and our immune system to our hormone balance and our ability to access the power of self-healing to bring happiness into our lives and our outer circumstances.

True limitations are found only in the head, or, more precisely, in the mind. Otherwise, a wide variety of riches and endless possibilities lie within each of us. And the wonderful thing is, this is not just "a thing you have to *believe* in" or pure conjecture; science has validated the truth of this. Studies show that we not only influence our individual lives through our beliefs, we also influence our entire environment. And through our thoughts and emotions, we have the ability to make any kind of change in our lives we wish to make. With new beliefs replacing old, outmoded beliefs that no longer support us, we can change ourselves all the way down to our DNA, stimulate our self-healing powers, breathe luck and joy into our lives, and achieve virtually anything we perceive as being possible.

Things are impossible only if we believe they are impossible. The reverse is also true.

Maybe you are utterly convinced that the very idea of unrestricted possibilities is impossible. That, then, is your belief. There is nothing right or wrong about it, nothing good or bad—it simply is your belief. And your life will arrange itself and develop according to this belief.

But what if your beliefs and your worldview are based on incorrect facts and information?

The most recent scientific discoveries prove unequivocally that through our thoughts, feelings, and beliefs, we are capable of anything. And it is precisely these emotionally supported and stored beliefs that build a powerful resonance. Everything—everything in this world—that resonates with these beliefs is grasped by the vibrational patterns of these beliefs, and the result is a resonance field.

So, knowing that you are creating your own reality, the question is: what type of resonance field are you creating?

If you could be who you always wanted to be, who would you be? And what prevents you from becoming that?

PART 1

Discoveries That Changed Our World

The Law of Resonance

- There is an energy field that connects everything.

- This energy field communicates with our resonance field.

- We build our resonance field through the "speech" of feelings and through the energy of thoughts, but most of all through our deeply felt beliefs.

- We send these beliefs out with our heart, our DNA, and the power of our thoughts.

- Neither distance nor time plays a role in governing our resonance field.

- We are connected with everything and everyone through the law of resonance.

- Who or what steps into resonance with us cannot help but react to us.

- Everything and everyone in resonance with us is inevitably drawn into our life.

- Likewise, we are inevitably drawn into the resonance fields of others if we oscillate in the same manner as them.

Each Era Has Its Own Breakthroughs

What if all your beliefs are based on incorrect facts and information?

Pierre Franckh

You might find the first part of this book a bit challenging to your understanding of reality because it is based on a different way of viewing reality. Don't be surprised if you find yourself becoming defensive because you either do not believe, do not understand, or simply cannot accept the following. Whatever reaction you experience is basically insignificant. What *is* important is that everything you are about to read is scientifically proven fact.

You might find it reassuring that it took me a long time to internalize this type of thinking. So try not to become impatient with yourself if your mind has a hard time accepting the scientifically proven findings.

Each era has had its own advancements and has touched new frontiers. We find ourselves today at such a new frontier.

Ultimately, it is one of the most satisfying experiences to be able to participate in a change of consciousness such as what is occurring now. The new realizations that are coming forward with this change clearly explain the law of resonance and how our *successful wishing concept* can be effective in helping us achieve our desired goals.

What Is Resonance, Really?

Resonantia = Reverberation
Resonance = Echo, Reverberation, Resonate

Through the law of resonance we come to understand how everything in the universe communicates with every other thing through vibrations. All objects and living things in our known world possess a unique oscillation. This includes the microcosm of the organs and cells of the human body, and extends beyond our individual bodies to include all matter. When we examine matter in terms of its own resonance energies, we discover that different objects generally oscillate with different frequencies, while some oscillate with the same or similar frequencies.

We know this from playing piano: when we hit a key on the piano, all the strings that resonate with the struck note—i.e., those that recognize the struck note and harmonize with it—are likewise brought into oscillation. The notes could be higher or deeper, but as long as they are in resonance, they are brought into oscillation.

Other people, things, or events cannot escape the resonance field that is generated within us if those persons, things, or events resonate with our generated frequency.

Much like the various strings of a piano that resonate with a struck note, other people, things, or events cannot refuse to resonate with us if they are in the same resonance field as we are. They must and will react according to this resonance.

What is the advantage when others with our same energy are brought into resonance with us? Here comes the second fundamental truth about the law of resonance: **Like attracts like.**

Everything that resonates with us is inevitably drawn into our life. It is not necessarily always "positive" for us.

For example, an opera singer, through the sheer vibration of her voice, can shatter a glass with a note. Her voice, sounding the note, essentially leads energy through the room toward the glass. When the transferred energy has the same resonance as the glass and the same natural frequency as the molecular structure of the glass, the stress on it can become so great that it shatters the glass. Resonance can be so strong that it destroys matter.

Of course we do not "burst" like glass, but similarly the so-called negative vibration that we carry within us can bring the negative parts of ourselves into motion. This can bring people or things we actually do not want in our life or bring uncomfortable, possibly even shocking, events into our life. Conversely, positive vibration or resonance will bring the people, things, or events that carry the same positive "charge" or resonance into our lives.

That is why it is important for us to know exactly which unique frequency we find ourselves in and which resonance field we are consciously or unconsciously creating.

How Do We Really Send Out Our Wishes?

It is harder to crack prejudice than an atom.
ALBERT EINSTEIN

Since the dawn of human civilization, the heart has been considered the strongest symbol of love, the center of our feelings. Then modern science and Western medicine emerged and, wanting to make us the wiser, declared that the heart is basically only a pump that circulates blood through the body.

Although we cannot offer double-blind studies or other scientific proof to substantiate that the heart is the center of our emotions, humankind has continued to hold to the persistent belief that it is. A multitude of phrases come to mind: "She died of a broken heart," or "Keep love in your heart," or "The hate in his heart consumed him," or "The best things are felt with the heart"—and many more similar statements. In Latin, the word *courage* means "to possess a heart." Again, the word *heart* is related to one's feelings, in this case, a sense of mental and emotional strength.

How true this connection between the heart and feelings is, and how wrong science has been shown to be as a result of some surprising discoveries made in 1993, which never came to the attention of the mainstream. But then, science sometimes finds it difficult to admit its mistakes.

The Institute of HeartMath, founded in 1991 and known worldwide for its groundbreaking discoveries,* led groundbreaking research

---

*Those who want to read about the detailed discoveries of the Institute of HeartMath can do so at their website, www.heartmath.org.

in emotional psychology and the heart-mind connection. In 1993, researchers at HeartMath undertook a study of the power of feelings over the physical human body. Specifically, they focused on that region of the body widely considered the origin of our emotions: the heart.

Right from the start of this investigation, researchers made a remarkable observation: the heart is surrounded by a large, powerful energy field, approximately two-and-a-half meters in diameter. In fact, the scientists were downright stunned that this was not discovered earlier.

The heart generates an energy field approximately ten feet out from the body.

One needs to imagine this: the heart generates an energy field that is greater in size than the energy field of the brain. Until this discovery, science had asserted that the brain, with all its electromagnetic impulses, possesses the largest "sender radius" of any organ in the body. But then a bioenergetic field significantly larger than that of the brain was found—the heart—and it possesses such strength that it extends far beyond the physical body. In fact, it is widely accepted that the energy field radiating from the heart has the largest radius that can be measured emanating from the human body, although current measuring devices are somewhat inadequate in pinpointing precisely the outer edges of the heart's energetic field.

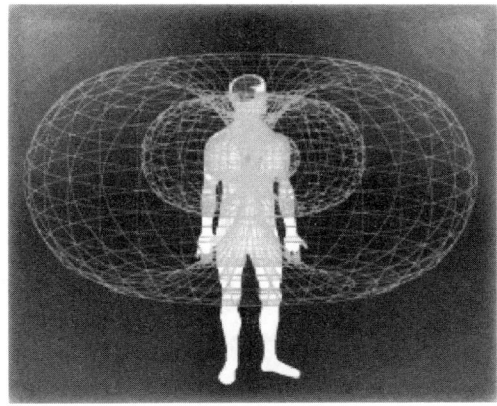

Representation of the electromagnetic field of the human heart. This field not only fills each cell of the body, it also includes regions outside of the body. (From Gregg Braden's The Divine Matrix, *by permission from KOHA Publishing)*

After this surprising discovery was made, the question arose as to exactly what sense organs are involved in this energy field that surrounds the heart. The results of this investigation were as remarkable as they are surprising: the electromagnetic field that surrounds the heart communicates with every organ in the body.

The electric and magnetic fields that originate from the heart communicate with all the organs of the body.

Notably, there is a signalized connection between the heart and brain that creates hormones, endorphins, and other chemicals in the body.

The brain does not work independently, rather it contains the signals that emanate from the heart.

It is the heart that distributes all information! But in what way does it "communicate" with the brain and the other organs?

In other experimental series it has been found that all information is transmitted by means of the *emotions.* All information is contained within our emotions, and it is the heart that transmits the information stored in the emotions to the brain and the organs.

But that is not all. As one delves further into this research, one discovers that the electrical and magnetic fields sent from the heart not only build up our emotions, they also receive power through other significant sources—namely, through our ***convictions***, through the things we believe deep down within ourselves! And it is by these beliefs that we direct our lives.

The echoes of this age-old truth are found in many expressions: "He defends his cause with the utmost belief," "It is her heart's desire," "He wished with all his heart," and of course "To speak the language of the heart." And so all of this finds itself as energy going out of the heart, carried by means of the highest sending power—electromagnetic energy—to the entire body. And this energy is not only transmitted to the brain and the other organs of the body, it is transmitted out into the whole wide world.

Our heart serves as a kind of intermediary that translates our beliefs and feelings into corresponding electrical and magnetic vibrations and waves.

And so our electromagnetic waves are not limited to our individual bodies; this energy reaches out far and wide into the environment and interacts with everything that surrounds us.

To summarize, our heart translates all our beliefs, all our perceptions, and all our emotions into a different kind of *language,* a coded language, of vibrations and waves, which it then sends out to all the systems of the body and beyond, into the environment.

Our beliefs are embedded in electrical and magnetic waves, which our heart sends out as we interact with the physical world.

Research from the Institute of HeartMath shows us exactly how large this sent energy is:

- The electrical power of the heart signal (EKG) is up to sixty times stronger than the electrical signal of the brain (EEG).
- The magnetic field of the heart is 5,000 times stronger than that of the brain.

We essentially send out considerably more energy through the heart than we do through the brain.

Why is it so important for us to know this? Simply because once we realize the power of the heart to affect our lives and our world, we can begin to understand why many of our wishes are fulfilled easily, while other wishes and desires do not manifest in our lives despite all our efforts. For example, we may constantly say affirmations and visualize our wishes being fulfilled, but if we do this *without* a corresponding *strong emotional component* that involves believing, then we are thwarting all our efforts. The brain may be sending its electromagnetic waves full force, but the true center of our feelings—the heart—will be acting according to our true beliefs, which in many cases center on our doubts and fears. And the magnetic field of the heart, being 5,000 times stronger than that of the head, broadcasts this out into our world. The results are undeniable— we will manifest what our heart is telling us about our beliefs. We can truly accomplish only those things that we believe in the *deepest recesses of our hearts*.

When we strengthen our convictions with the power of our emotions, the emitted energy is great. If we are sad or depressed, or if we find ourselves in an emotional hole, the power of the emitted somber emotions that are sent through the heart region is always considerably stronger than any wish we send through our understanding, through the mind. That is why the prophets, the wise ones, and world teachers of past and present times have said over and over again that we need to learn "to see with the heart."

We can change the world through our heart.

What most spiritual masters and texts have taught is this: that through our faith we can move mountains, becoming deeper and outside all scientific conception.

> Jesus replied, "Truly I tell you, if you have faith and do not doubt, not only can you do what was done to the fig tree, but also you can say to this mountain, 'Go, throw yourself into the sea,' and it will be

done. If you believe, you will receive whatever you ask for in prayer."
(Matthew 21:21–22)

This biblical admonition has now been proven by science. Only the strongest belief has the power to create something new in our world.

That which we believe is realized because having been sent through the heart, it has the strongest measureable energy.

To summarize:

- The heart signals to the brain which hormones, endorphins, and other chemicals should be produced in the body.
- The heart is the strongest transmitter in the body. It produces the strongest magnetic and electrical impulses, which we have at our disposal at any given time.
- The magnetic and electrical waves that come from the heart are created through our feelings and beliefs. Whether these waves are positive or negative is irrelevant; they are sent out with the same amount of power into the world.
- The heart is a type of transmitter that translates our deeply held beliefs into another language, a coded language of energetic waves that is sent out with tremendous energy.
- This means that our beliefs are sent out and are attracted to an equal energy, an energy that matches these beliefs, according to the law of resonance.
- Like attracts like. Everything that resonates with our energy is realized in our lives. In short, our beliefs manifest in our lives.

Therefore, this is the most important thing we need to know about wishing:

- Whatever you want, bring it from the intellectual level to the heart region.
- If you would like to see your wishes fulfilled, you must be utterly convinced that your dreams can be realized.
- To see your wishes fulfilled, you should put yourself in a happy mood first.

When our awareness is specifically aligned with our heart and our emotions, we can step into resonance with the things we want to realize in our lives.

In our world, only that which we *truly believe from our heart* can come to fruition. This is especially true of the things we believe *about ourselves*. The opinion we hold of ourselves determines our experiences. Naturally, this means that we must realize that all the power and strength we need to realize our wishes or our visions comes from within us and not from the outside. That is because the outer world is but a mirror of our inner consciousness.

Only when we align our consciousness with our target can we resonate with the things we want to bring into reality.

How Can Our Beliefs Change the Outside World?

What lies behind us and what lies ahead of us are tiny matters compared to what lives within us.

HENRY DAVID THOREAU

Until recently, we were convinced that in our world everything is separated from all. We believed that two separate things didn't have any influence on each other. Logically, we were taught, we have to look at ourselves as separate from the other. This inevitably generated feelings of isolation and loneliness within us. It seemed that things and events happened *randomly*. We grew up with this in mind.

There is the individualized I/me, and there is the rest of the world, the you/them. This dualistic view is so pervasive that we do not even question it, even though it does not bring us any sense of emotional well-being. Certainly it reaffirms the opinions we hold about our life.

It has only been in the last few years that the view of modern science on this subject has completely changed. Today we know the exact opposite to be the case: we are not separated from one another; everything is interconnected and thus every thing influences every other thing simultaneously. Because this knowledge is important for understanding how our wishing energies work, I will expand on this.

The change in perspective, from dualism to interdependence, began to emerge in the scientific community in 1995, with research at the Russian Academy of Sciences under the direction of quantum physicist

Vladimir Poponin and Peter Gariaev, the "father of wave genetics." The outcome of their experiments showed something very remarkable—so remarkable that they were repeated in the United States and the results were published here.

Poponin and Gariaev wanted to research the contents of DNA based on light particles—so-called photons (light energy). In this series of investigations that has since been dubbed "the DNA phantom effect," they removed the photons with a pipe or valve of air and a vacuum. Heretofore, the conventional view of science was that a void cannot truly be created in a vacuum. In this experiment, in each space, some photons remained, which could be measured exactly with specialized instruments. In the beginning, everything occurred as expected. The photons separated in the vacuum pipe in a relatively unorganized manner. In the next phase of the experiment, a sample of human DNA was contained in the pipe. And then something very surprising happened: the light particles organized themselves in a different order. In other words, the DNA had a direct impact on the photons. The photons in the pipe formed themselves into an orderly pattern, as if through an invisible power. With this discovery, one thing became clear:

Human DNA has a direct effect on the physical world.

Nothing like this had ever been observed in conventional physics. Moreover, an investigation such as this one had never been designed in the context of conventional physics.

The photons also performed some functions that at first had no explanation. This was exciting enough on its own, yet what followed was simply revolutionary. When the DNA was removed from the pipe again, it was assumed that that the sent order of the photons would dissolve and they would go back to the unordered distribution in the room. But the opposite occurred: the photons behaved as if the DNA were still present and remained in their ordered division.

Since then, other researchers recreated this study many times. They double-checked their instruments and double-checked the numbers to make sure that none of the DNA remained in the pipe. But all the studies led the experimenters to the same conclusion: the photons and DNA were still bound to one another, even though they were physically separated from one another. They seemed to be connected to one another in a field that quantum physics calls *the quantum field*.

The so-called empty space is not really empty—it is a field through which billions of information waves move and interconnect.

This experiment is very essential to the understanding of the law of resonance. The series of experiments support and underpin the theory, the existence of a quantum field in which everything is interconnected. What made this experiment so special is the fact that this energy was scientifically demonstrated as being observable.

Since its discovery, there have been many names for this energy field that connects everything: *quantum field, godly matrix, the field, quantum hologram, the source field, the torsion field*. What is special about this energetic field is that it is not similar to any other form of energy we know about.

This energetic field that seems to function like a tightly woven net creates a type of bridge between the inner and outer worlds.

In the same way that sound uses the air, vibrating the air molecules, which then push on other air molecules and so on, until a wave is created, our released energy that holds our beliefs and thoughts also uses a medium—the quantum field—in order to be carried around the world.

This energy field makes it possible to be connected with everyone and everything, either consciously or unconsciously.

The distance between us and the recipient of our thoughts and beliefs is irrelevant; it doesn't matter whether the person is our husband, our neighbor, or someone on the other side of the world. The generated and sent energy always finds the right person, even if that person isn't consciously aware of this sent energy.

Sabine, for example, discovered exactly this; through her wishing energy, she found her connection with a longed-for partner, although at the time he was completely oblivious of her or her sent energy.

> *Hello Pierre,*
> *I was at your seminar in Frankfurt in June of this year, and what happened since then can barely be put into words. It has been a time of almost effortless fulfillment of my many wishes. But only one wish, which lay most heavily in my heart—namely, to find a life partner again—seemingly fired with no resonance. So I decided to follow your advice and first be thankful and happy with all that I have.*
>
> *At the beginning of August, I went on vacation with both of my children. And these phenomena repeated themselves. Even forgotten wishes were fulfilled. With that, I drew in people who were nice and partially interested in my business. Only I was still waiting for "the man."*
>
> *So I decided to put a small turbo into my affirmations. Each morning, I did half an hour of concentrated belief sets while working on the cross-trainer.*
>
> *And suddenly, "he" stood in front of me. While in a disco on my way to the bar, this man almost ran me over to make it clear to me that he absolutely had to get to know me better. He had to get some liquid courage to do this, but he had waited for days for the perfect opportunity to talk to me. He had no idea why he felt compelled to talk*

to me, but he could not do anything else, since he felt that something had driven him to it.

This pickup was so unusual; there was something to it—an unbelievable attraction was there from the first moment. It turns out that this man fulfills the collective wishes of my wish list and that we have so many similarities that it is almost uncanny. In the meantime, we could not imagine how we ever lived without each other.

Moreover, it seems that his readiness developed at the same time as my availability. I added this wish to my wish list, thanks to you, after your seminar.

I have waited so long for the fulfillment of this heart wish. And now, he is here.

You and I will see each other again at one of your seminars, but the next time I will be bringing someone with me . . .

Warmest wishes,
Sabine

Why wait any longer for the fulfillment of your heart's wishes, when you can be energetically clear and active in achieving your goals?

Sabine *simply began* to trust the power of her thoughts and steadfastly held to the beliefs of her heart. We too can be instantly connected to anything and anyone through the quantum field. We simply need to just do it.

*The law of resonance always says "yes."
It always confirms your beliefs.
It does not contradict you.*

*If you believe
That your life does not make sense
Or have a lot of deep meaning
Then this will be confirmed.*

*If you believe that you are ready
For a deep, true love connection,
Money, or inner and outer fortune;
If you believe that your life possesses
A deep, encompassing meaning,
This is exactly what will occur in your life.
When the law of resonance is followed,
Nothing else can develop in your life.*

*The energy is essentially indifferent,
Whether it is of a moral high quality or reprehensible,
Whether it is useful to you
Or whether it hinders you in your life.*

*The energy does not ask about morals
And does not evaluate.
Energy will always react
To your continuously sent impulses.*

Can We Impact Our Cells through the Power of Thought?

Facts do not cease to exist because they are ignored.
ALDOUS HUXLEY

We have known for a long time that our feelings have strong effects on our body.

This too is well documented in science and in medical circles: DNA supposedly is unchangeable. But is it really?

Between 1992 and 1995, the Institute of HeartMath (IHM) researched the effects of pure feelings on our DNA. Cell biologist Glen Rein and IHM research director Rollin McCraty used human DNA in these investigations. The DNA was isolated in a beaker, then exposed to strong, powerful emotions. To accomplish this, test subjects held a DNA sample in a test tube in their hands, and then the subjects were directed to use different spiritual and emotional techniques, such as the pacification of the spirit, concentration of positive feelings, and focusing on the regions of the heart.

The results were very impressive and not to be ignored. Individuals capable of generating high ratios of heart coherence—defined as a state marked by more ordered, or coherent, interactions between the body's various systems, including more ordered heart rhythms—were able to alter DNA conformation according to their intention, Rein and McCraty found. The control group participants showed low ratios of heart coherence and were unable to intentionally alter the conformation of DNA. Generally, participants with the highest levels of coherence affected the samples the most.

In short, the experiment showed that test subjects with high levels of heart coherence were able to affect the DNA molecules in the beaker simply through their emotions.

Human feelings affect the formation of DNA.

This may be hard to accept, since we have been taught that DNA is unchangeable. We are born with it and nothing—except possibly difficult, extensive surgery—can affect it or change it in any way, especially if it has been separated from the body from which it comes. And yet we have come to realize that DNA *is* absolutely changeable and in fact reacts to very subtle energetic vibrations.

The Institute of HeartMath went one step further in their research in this area and examined the reactions of human placenta DNA, the purest form of DNA.*

To do this, twenty-eight placenta DNA were divided into twenty-eight beakers and were handed to each of the twenty-eight qualified researchers, who were trained to produce very strong feelings.

The results of this study also proved that DNA changes its form based on the feelings emitted by each of the researchers. For example, when the researchers felt approval, love, or drunkenness, the DNA responded with relaxation; the DNA strings opened up, which means they became longer. In contrast, when the researchers felt frustration, anxiety, anger, or stress, the DNA strings became shorter and shut off many of its codes! In other words, the DNA reacted to negative feelings by pulling itself together.

Now we can understand why negative feelings simply *shut us off* from the outside world. When we are angry or in a bad mood, we feel

---

*The test results were published in the article "Local and nonlocal effects of coherent heart frequencies on conformational changes of DNA," in *Proceedings of the Joint USPA/IAPR Psychotronics Conference,* Milwaukee, Wisconsin, 1993.

isolated and cut off from the flow of life, and the reality is, we *are* in fact cut off. However, we also cut ourselves off!

As shown by these experiments, the disconnected, shut-down DNA code can be immediately repaired when we feel joy, appreciation, thankfulness, and love. In this way, the DNA codes are changed, as if we were switching a darkened lamp back on.

Incidentally, the changes that were measured in the DNA in this series of experiments were considerably larger and stronger than those created by electromagnets.

People who are deeply in love can change the formation of their DNA.

Then, when this same research was conducted with HIV patients, it was discovered that in those in whom feelings of recognition, gratitude, and love were present, the person's immune resistance increased by a factor of 300,000 over those who had no access to these feelings.

One can suppose that the key to good health lies in this discovery. Therefore, it seems quite advisable to practice feeling joy, love, gratitude, and respect, because we can thereby increase our resistance 300,000-fold. The conclusion is obvious: when we maintain positive thoughts, we can prevent many different types of illnesses, because these positive thoughts strengthen our immune system.

But can we do even more? Can we can go back and actually find our foundation of good health? There are countless success stories that prove that this is so, such as that of Sandra:

> *Dear Pierre,*
> *For me, wishing has mostly fallen to the back of my mind. Here is an example dealing with health:*
> *I suffered from an eating disorder called bulimia. I felt that I needed to force myself to eat an endless amount of food, only to finally*

throw it up so I did not get fat. For years, I had a double life that no one knew about. On the outside, it seemed that I was doing well, but on the inside I felt ashamed.

One day, someone recommended a book about positive thinking, which is nothing other than "successful wishing." I was immediately enthusiastic because I had thought that I was the only one who could change this situation. So I began with positive affirmations and auto suggestions. In addition, I always imagined how I would get compliments from others on being trim and healthy and having a positive aura and figure. It took a while, but that was the only way I achieved success. My life changed from the ground up. Everything came as it should. Next, I had to get accustomed to eating regularly without vomiting. In this I was not as successful, until I banned the scale from my life. Then I lost nearly twenty pounds in less than a year, and this weight loss was lasting. Even today, I am at my ideal weight of approximately 120 pounds, and I can eat whatever I want without weight gain. With that, my bulimia is a thing of the past, and without any outside help I can maintain my ideal weight.

Warmest wishes,
Sandra

I myself have experienced the power of self-healing in my own life—healing that, according to classical medicine, should never have occurred.

I was just twenty years old when I started to suffer from strong back pain. I was immediately sent to a medical genius in this field. The diagnosis following my first examination was devastating: I suffered from ankylosing spondylitis, a form of spondyloarthritis, a chronic, inflammatory arthritis in which immune mechanisms are thought to have a key role. It mainly affects joints in the spine and the sacroiliac joint in the pelvis, and can cause eventual fusion of the spine. In a few years, my back would completely stiffen. There was no doubt about this diagnosis, I was told; all the lab tests had confirmed it.

The doctor who saw me said that a week later, after the first shock of this diagnosis had settled, we should meet to discuss further therapeutic steps. It was clear: this illness was incurable. The life of an actor that I had dreamed of was clearly no longer possible. In a very short amount of time, my spinal cord would stiffen so much that it would be like one single bone. And not only that, my back would become so warped that I could only see the floor.

One week passed—one of the most difficult weeks of my life. At that time, I did not know anything about "successful wishing" or quantum physics. I did not know that one could change one's DNA simply through one's thoughts. I only knew this: "I am healthy. My back is wonderful. I love my back. I am powerful and strong. I am lively and agile." I even went ahead and booked tennis training hours for a year in advance—not because I was desperate, but because I was so happy to be alive and because I knew how good it was to be healthy. For me, there was no doubt: my life had just begun! I laughed as if a funny joke had been told. I allowed myself to wake up in the morning and feel the life in me, feel that I was healthy. I concentrated day and night, every single minute, even every single second, on the wonder that was being unlocked within me. This, by the way, was very surprising for those around me. No one could understand my good mood, my happiness, my love for anything and anyone. I danced, I sang, I was connected, I was *one*. I was full of gratitude and appreciation for my present life.

Then came more doctor's visits, further examinations, and lab tests. Even my doctor wondered about his smiling and cheerful patient. And then the bomb burst. Unexplainable! Unbelievable! This had never happened before! An impossible change! The stuttering doctor stood in front of me. The rheumatic inflammation, the spinal column, the weakened sacroiliac joint—it was all so clear . . . and yet, "We need to examine it once more," he said. But three days and more tests later, the doctor, an eminent authority, had no explanation for what had happened to me. Disappeared. Gone. Nothing to see. Nothing else to prove. The erythrocyte sedimentation rate was completely normal. No

HLA-B27, no spondylitis. He was sorry about his earlier diagnosis—no, he was naturally very happy, but bewildered . . .

I actually believe that this man would have rather had everything go through the "normal" horrible process he had predicted, as long as *his* life continued along its regulated path. But all he could offer me was a handshake and a disbelieving head shake. I still remember how I comforted him and took his arm and reassured him, how I kissed the medical assistant—and then began playing tennis.

Today I know what happened back then. And I continue to be deeply touched by this gift from myself to myself, as a result of the power of my own thinking.

From where did I get my self-assurance? Well, I was fortunate: I always believed my intuition and my inner truth, rather than unquestioningly following the opinions of others. Looking back on it now, I know I received an unbelievable amount of additional confidence in my own inner truth, which has remained with me up to the present.

This experience, as well as others, has convinced me that we all have our health in our own hands, more than we know. This is my firm belief.

And what do you believe? Your belief is the decisive power in any situation. Decide for yourself what you want and point yourself in that direction.

*The strongest energy
that we have at our disposal is
Love.
Fall in love with your wishes!
This will produce the greatest
amount of positive energy.*

Can We Impact Far Distant DNA through the Power of Thought?

Problems can never be solved with the same way of thinking through which they were created.

ALBERT EINSTEIN

The knowledge that we can change our DNA through our thoughts deeply shook the fundamental understanding of many scientists. Does the world function differently than previously thought?

In the early 1990s, scientists under contract with the U.S. Army undertook research to see if our feelings have an influence on living cells when they are located somewhere else, for example, some distance away from the body. This would seem doubtful because based on what we knew up to that point this is physically not possible. People also did not believe that, for example, organs, bones, skin, or tissues can remain connected with a person from whom they have been taken.

In 1993, an article was published by Glen Rein and Rollin McCraty that reported on these experiments by the U.S. Army.* Studies were conducted to see if a connection between DNA and the feelings of subjects could be detected. DNA and tissue samples from the mouth of the test subjects were taken. These were isolated and brought to another part

---

*See "Local and Non-Local Effects of Coherent Heart Frequencies on Conformational Changes of DNA," Glen Rein, Ph.D., and Rollin McCraty, Ph.D., http://appreciativeinquiry.case.edu/uploads/HeartMath%20article.pdf.

of the same building where the research was being conducted. Using devices that were made especially for this purpose, scientists wanted to see if the DNA of the donor showed any reaction to that person's emotions, even though it had been separated from the donor.

To generate certain feelings in the test subjects, various images were shown to them, such as erotic photos, violent scenes, comedy scenes, etc. The full breadth of human feelings was covered in these images in order to send the subjects into emotional highs and lows.

It should be noted that most of the scientists who undertook this study doubted that there would be any observable effects on DNA. Yet while the test subjects experienced their feelings, the researchers measured the electrical reactions of the DNA and discovered that the DNA behaved exactly as if it were still in the bodies of the donors.

Gregg Braden writes about these experiments in his book *The Divine Matrix*. He recounts how subsequent to these initial investigations, Dr. Cleve Backster, an interrogation specialist for the CIA, took these studies further by extending the distance between the donor and DNA to as much as 350 miles.

In Backster's DNA study, the delay time between the transmission of feelings and the reaction of the DNA was measured by means of an atomic clock, and it was determined that the reaction of the DNA followed immediately. Yes, immediately—it came virtually instantaneously, with no delay. The reaction was found to be exactly as fast as it would have been if the DNA had still been in the body of the donor.

What we feel, think, and believe is collected by our DNA in the same millionth of a second.

It is irrelevant if the resonant DNA is right next to us or is halfway around the world; the fact that emotion can alter DNA is not limited by time or distance. The net result suggests that we direct a force within us that operates in a realm that is free from the limits of physics as we have known it. This energy field, or matrix, or quantum field—whichever name we choose—is the medium in which all our feelings and thoughts move and operate, and not only with the speed of light, but even faster than that.

Dr. Jeffrey Thompson, a pioneer in the field of brainwave entrainment frequencies incorporated into musical sound tracks to facilitate mind-body healing, has done extensive research on how the quantum field operates.* Thompson formulated his discoveries as follows.

There is no single point where the body ends and where it begins.

The Pavlov Institute of Psychology in Moscow made a further discovery with regard to the time-free nature of the quantum field and the power of the emotions. In one study, six young rats were removed from their mother and taken to six different places around the world. The removal of her young generated in the mother feelings of panic, fear, or joy. And as if to confirm the discoveries of U.S. researchers, the Russian scientists found that the young rats, although located in different places around the world, reacted to the feelings of their mother immediately.

These discoveries represent milestones in how science has come to understand how human emotions operate and the construction and work of human cells. For me, this is yet another explanation for why *successful wishing* functions so well.

At the beginning of this book we learned that DNA has a strong effect on its environment and leaves behind a lasting and permanent impression. The previous two chapters reveal that we can influence our DNA through our thoughts and feelings. That we can influence our own DNA—and through the quantum field, outside of the constraints of

---

*For more on Dr. Thompson's work see www.neuroacoustic.com/articles.html.

time and distance, we can connect to everything else in the world, and in so doing we can draw anything we want into our lives—is a fantastic notion to many people. In fact, whether we acknowledge it or not, we are already doing precisely that. This is the essence of the law of resonance.

Everything that we possess in our inner world we also encounter in the outer world.

That which we encounter in the outer world has an inner origin, and this is our thoughts. Therefore, if we want to get the results we wish for, we should begin to observe our minds.

- Everything we think, feel, or say intensifies our resonance field. This is why every thought of loss furthers loss, and why every belief concerning gain brings about gain. Knowing this, anything we want to change in our outer world can only be changed first through our thoughts.
- Remember to use your innermost creativity to manifest that which you desire, and use it to become more aware of your well-being and to bring about the well-being of all.

Sabine, whom we heard from earlier, wrote to me about another experience she had with wishful thinking, which is a good example of how powerfully our thinking can affect others.

Hello, dear Pierre,
It is unbelievable what has happened to me as a result of wishful thinking.

Jannik, our son, is a very lively small boy of three years. At the beginning of the year, he suffered a fall "in the heat of the moment" in which he hit his upper lip. He was bleeding, and I, like any mother would, doctored him and assured him that he didn't really hurt anything, as every mother does. Jannik calmed down quickly and the wound immediately stopped bleeding. Because of this, I knew he was okay.

Weeks later, I noticed that his left front tooth was clearly darker than the rest of his bright white milk teeth. I arranged an appointment with our dentist. He examined the boy, took an X-ray, and asked me if Jannik had fallen on his lip any time in the last several weeks. I immediately thought of his accident. The doctor explained that according to the X-ray, the nerve that supports the tooth was severed during his fall and that the tooth would become darker and darker. But I should not be concerned, he said, because as soon as his permanent teeth came, this tooth would be as white as the others again. But this would take another few years because as everyone knows, the first permanent teeth don't come until the age of five or six.

This seemed too long to wait, so I began to wish . . . I wished that the color of the tooth would change back to bright white. For any rational person this might sound like an absurdity. Nevertheless, a change could be clearly observed. Week after week, it was easy to see a clear change. The tooth got its previous color back! After approximately four weeks, the tooth was completely white again. I called the dentist to ask if he had ever seen such a case in his years of practice. He could not believe it. For him, this was unexplainable.

Today, six months after the fall, it is impossible to see a change. The previously dark tooth is now exactly as white as the others.

Sabine

How Do Our Wishes Become Reality?

I wanted to change the world, but I found that the only thing one can be sure of changing is oneself.
                                                ALDOUS HUXLEY

How do all our wishes and longings arrive at the place where they should be? And how does what we wish for come back to us? How do the receivers of our wishes know where they should even find us? What part of our body receives the information and passes it on to our consciousness? How can we integrate this new awareness into our daily life?

Our DNA plays a large part in the answers to these questions. It is the carrier of our genetic code. Since the discovery of DNA, people have believed—and I also learned this in school—that DNA is exclusively occupied, with the help of our genetic code, with making protein bodies in the inner portions of the cell. Surprisingly, however, almost 90 percent of our DNA is not needed for protein synthesis, but rather is essentially used for *communication*. The previously mentioned Russian scientists Vladimir Poponin and Peter Gariaev proved this: DNA does more than what was initially thought.

DNA communicates with its environment.

Poponin and Gariaev showed that DNA is really predestined to work as a sender and receiver. And not only that. These scientists proved that our DNA communicates not only with us, it also communicates with the DNA of other people. Meanwhile, we know that our DNA is connected with everything that is.

With this, we come to another very surprising discovery in science: communication between our DNA and the DNA of other people and other living things is achieved completely differently than we initially thought. It occurs on a higher dimension, outside of space and time. Therefore, the term *hyperspace* has been coined.

The astonishing thing about these special exchanges of information is that they occur without any time or space restrictions. Neither distance nor time presents a problem for the smooth exchange of information. There is not even a hint of a time delay.

To accomplish this, DNA uses the special energy channels known as *worm holes,* which exist in hyperspace. Albert Einstein and Nathan Rosen were the first to describe worm holes, as early as 1935. The concept of the worm hole was defined as being two sides of the same space that are connected by means of a tunnel.

Because of these worm holes, the distance between us and the person affected by our resonance frequency is irrelevant. The person could be lying in the bed next to us or be halfway around the world. They can be awake or asleep. All information that we send out is sent in hyperspace through these energetic tunnels and immediately comes to its target, to be received and used by the local DNA.

Incidentally, it should be mentioned that this energy is not only received by DNA, it is also saved by it. Therefore, DNA also serves as an information recorder. In other words, we have a huge data bank in our body.

So now the important question: How does wish fulfillment find us? How does this energy that steps into our resonance even locate us to begin with? Ultimately, there are billions of different DNAs; each of them sends and receives. So how does the cosmos purposefully grant us our wishes?

First, we should know that we are constantly broadcasting, constantly programming our resonance field with our thoughts, whether positive or negative. As long as we maintain our wishes and visions—or our fears and misgivings—our resonance field draws it in.

Next, each of us possesses a unique genetic code, which we have already heard of in the context of forensic and paternity testing. This means the DNA of each person is as unique as that person's fingerprint. The code leaves an unchangeable genetic fingerprint. And just like an ordinary fingerprint, the energetic fingerprint of our DNA leaves a clear imprint. The oscillation of this imprint is so precise that it always finds the appropriate solution for us, which is shown in Anka's example, below. Those who know my book *Wünsch es dir einfach—aber richtig* (Simply Wish It—But Correctly) have already encountered Anka's astonishing story of how she wished her beloved son back to health. Here is another, equally astounding story from her:

Dear Pierre,
First of all, thank you for the two copies of your book Wünsch es dir einfach–aber richtig *(Simply Wish It—But Correctly). I can finally report to you that my son is perfectly healthy, and no trace of neurodermitis can be found.*

I was very depressed when I recently returned from vacation. We traveled through Italy in an RV for a few weeks. Toward the end of the vacation, we were robbed while we were asleep: money, papers, etc.—all gone.

When we noticed the robbery in the morning, we were stunned. My husband started pacing out of anger, and I was paralyzed because I simply could not believe what had happened. I had never been robbed before!

We found that we were apparently the only ones who had been robbed where we were parked. The question why? arose. Remarkably, my handbag with my cell phone and some cash had not been taken. It was still there in the vehicle, where it sat in the baby seat, undisturbed.

At that point I wished that everything would be okay. The luck ran its course.

We drove to the police station in our giant RV, where parking was out of the question. Kindly, someone reserved an extra place for us so we could park directly in front of the entrance, where we could drink our coffee in peace.

Although the station was packed, we were given priority, as we had our young son with us. The officer we spoke to gave us no hope that we would get our valuables back. When he asked us if we had been chloroformed, we realized just how lucky we had been. An anesthetizing gas could have had bad consequences for our then nine-month-old son.

We received a text message from some friends saying that they were coincidentally on vacation only a few miles away. On the way there, we noticed that our son's diaper bag had been stolen, including the pacifier, examination booklet, and his first pictures. And then the tears came. I was tired and hungry and just wished for a relaxing evening.

Our friends not only helped us with cash, they also invited us to a wonderful dinner. Our son quickly fell asleep in his car seat and we were finally able to relax.

Following our vacation, we got home and picked up our mail, at which point the two copies of your book immediately fell into my hands. At exactly this moment, I knew: we would get everything back that was really important. And after I read the first few pages of your book, which said that my letter, the story of my son, had been a sign for you to continue your writing, it was as if you had thrown the energy ball back to me.

In that moment my earlier question as to WHY the theft had occurred the way it did was answered: I realized my husband had used an incredible amount of energy before our vacation to find an appropriate theft insurance, which did not exist! But I do not have to tell you why my handbag remained undisturbed . . .

Shortly before the end of the year, I asked myself why my wish had not yet been fulfilled. I believe that my husband felt a bit of triumph because he still cannot truly believe in the success I have from my wishing. Some days later, exactly one day before my son's first birthday, I received a letter from the lost and found in Munich: all of our things, with the exception of the cash (which was not important to me) had been found! The most important—the first pictures of our son—were there! We even got back our precious briefcase.

Wishing works!

My husband asked me why I always write you about my wish successes. My answer, here also for you: gratitude!

I had not had much experience with "successful wishing," I had no preconceived notion of how the universe works to grant our wishes. How could I be thankful toward "something" (the universe) I had no concept of? But since my experience I have had an image of you in my mind's eye; I have met you with my gratitude. Today, I do know how the universe works to fulfill our wishes, and I am still endlessly grateful.

Sincerely,
Anka

*Tell me
and I will forget.
Show me
and I may remember.
Involve me
and I will understand.*

Lao-tzu

How Do Affirmations Work?

What we think, we become.
BUDDHA

Successful wishing is nothing other than a very appropriate way for us to reprogram our own code. Whether through positive affirmations, autogenous training in meditation, autosuggestion, or hypothetical visualization, our DNA can always receive and store this information. Remember, there are no boundaries. The only true boundaries are in our heads.

It is really the power of human belief that allows us to be that which we believe.

When there is a desire to be healthy, you do not deny the illness, instead you enable your self-healing powers. We mentally replace the vibration of illness by visualizing pictures of radiant health and generating the resonance of health. And so, rather than suggesting to our body that there is something wrong with it, we allow the energy of health, as generated through our emotionally held beliefs, to work in our body.

The body reacts to the smallest thought impulses. When we believe in our health, we can then induce the body to mobilize all its self-healing powers on all levels. What many ultimately believe to be a wonder once they have regained their health is really not a wonder at all, but rather a confirmation of how strong the power of thought really is. This is illustrated in Monika's letter:

Dear Pierre,

Because news of the strength of my wishing—and most of all, the success of my wishing—has spread through our small circle of acquaintances, a friend called me one day, completely distraught. She knew about all the success I had from wishing, but she had a hard time convincing herself that it really works. She therefore called on my help when she was in distress.

She told me a story about her mother, who in a matter of a few minutes went from being a halfway healthy person to a cripple. A blood clot that had formed on her spinal cord was pressing on the nerves that controlled the lower part of her musculoskeletal system. In short, the movement of her legs, and with that, her passion, running, had been brought to an end in a matter of minutes. She had been brought to the hospital by a medical helicopter. There, she was immediately operated on. The doctors tried everything to stop the worst. The operation lasted a long time.

After my friend's mother awakened from the anesthesia, tests were immediately performed to get a clear indication as to whether the operation had been successful. According to the attending physician, immediately after the operation movement in the legs should have been noticeable. But nothing happened, even days after the operation. The doctors confirmed that nothing positive had occurred and concurred that the patient would probably spend the rest of her life in a wheelchair. My friend was at the end of her rope as a result of this news concerning her mother.

I was naturally very upset about my friend's mother's condition; I took it all to heart. At first I did not know just what my friend wanted from me until she suddenly told me that she wanted me to wish for her mother to run again. After all, I had already achieved "medical" success with my son. I agreed with her and assured her that I would send my wish to the universe on that same day. I hung up the phone exactly at 3:30 p.m. Immediately thereafter I sent my wish out into the universe. And at 7:15 p.m. I received a text message from my friend. She had just returned from the hospital, where her mother had demonstrated that she could move her right foot!

I was not at all surprised, because I knew the power of wishing is endless . . .

Today, fourteen days after the paralysis occurred, my friend's mother is able to move her thigh some. She is now in rehabilitation and makes daily progress—against the expectations of all her doctors!

Monika

*Science cannot solve the ultimate
mysteries of nature.
And that is because
We ourselves are part of nature
And are therefore also part of the
mystery we are trying to solve.*

Max Planck

Can We Create a New Future through the Power of Thought?

> *People like us, who believe in physics, know that the distinction between past, present, and future is only a stubbornly persistent illusion.*
>
> ALBERT EINSTEIN

Can we influence our future through our thoughts? An unequivocal yes! We can do that, perhaps more so than we thought. Quantum physicists have discovered something very exciting, showing again that we can essentially completely change our lives at any time, and that we can make anything in life a reality.

As we now know, human beings transmit energies through their thoughts. All people do. We also know that energy with a certain oscillation attracts reciprocal energy. This is a logical effect, since we not only attract people and events toward us; to the same degree we are also attracted by other people and events. The requirement is that both energies resonate with each other similarly.

Now, quantum physicists have discovered that so-called quantum waves—for example, our thoughts and beliefs—work not only spatially but also temporally. They don't just spread out in space, they also spread out in time (as in the term *time waves*).

There are quantum waves that are communicated from the past toward the future that are known as *normal quantum waves*.

There are also energy waves—called *conjugate complex waves*—that are communicated from the future to the past! It is truly amazing, but this is just how it is.

The waves communicated toward the future are called propitional waves or "offer waves," and those communicated in return toward the past are called "echo waves." When two such waves meet, that is to say when an echo wave out of the future encounters an offer wave from the past that we have put out there, one of the two waves will shape the other, bringing about what is known as an *event probability*.*

According to quantum physics, the possibility that one event occurs results "from the meeting of an offer wave from the past and a 'matching' echo wave from the future," according to translator, writer, and editor Jörg Starkmuth, in his book *The Making of Reality: How Consciousness Creates the World*.

To understand better, let's compare this to how a fax device works. When we send a fax, the machine first establishes contact with another fax device and exchanges specific test signals. Only then, when both devices are on the same transmission waves, can the data be exchanged.

According to Jörg Starkmuth, ". . . the past and future communicate in similar ways with each another and through the meeting of matching signals create a concrete event of higher consciousness at the 'midpoint,' which is an experienced present. This means that not only does the past influence the future, but the future also influences the past!"

Our mind might have a hard time understanding this notion because we tend to see time in a linear fashion, and always from the past to the future. So now the opposite should be possible. John G. Cramer's

---

*This was discovered in 1980 by John G. Cramer, professor of physics at the University of Washington in Seattle, who studies retrocausality, or "backward causation." For Cramer's work go to www.npl.washington.edu/av/. For more on this topic in general see F. A. Wolf, "On the quantum physical theory of subjective antedating," *Journal of Theoretical Biology* 136, no. 1 (1989): 13–20; or go to www.absoluteastronomy.com/topics/Retrocausality. Since this time this has not only been confirmed and documented in following studies, but developed further.

theory has not only been confirmed multiple times, it has made specific paradoxes in quantum physics explicable. The truth of the nonlinearity of time means:

The future is not less real than the past.

The future already exists somewhere "out there." Otherwise, it could not send any waves into the past, which is our present. And your future already exists, right now, in this very moment. Nevertheless, it is not predetermined, because we each have different possibilities to choose from to create a different kind of future. In fact, science has shown that the possibility exists that we can choose our future.

So how does this work, that the future already exists? Does this mean it is already written? No, because *our future already exists in countless versions.*

Science has made significant progress since 1980, when John G. Cramer made his initial sensational discovery. The common belief then was that only one reality exists and, accordingly, that we all have only one possible future. Meanwhile, scientists were discovering that *there are different, parallel realities occurring at the same time.*

This therefore becomes truly interesting when we consider our *successful wishing energy* and our creative visualization.

- It is assumed that time does not function in a linear fashion as we perceive, but rather everything is happening at the same time.
- So the past is happening now in this moment, like the present and the future.

Our consciousness perceives only *one* time, the present. We do not know anything else. This is not surprising because our senses are limited. Our senses can only take in 8 percent of the entire light spectrum. Our senses cannot perceive 92 percent of our reality and that means that it simply doesn't exist for us. And yet it is there. But as we are our own "measuring instrument"—and a limited one—we cannot understand this and therefore refuse to believe it. Nevertheless, we are surrounded by a wealth of other energies, oscillations, waves, and information.

> *I know that I know nothing.*
> SOCRATES

This finding is so old and it has not changed much until today.

We actually do know what we do not know.

Even if most of reality remains completely closed to us, there is nevertheless a part of ourselves that works on the other side of our perception, even if we do not understand *how* it works.

This is quite similar to what happens with time. Also in terms of time, we can only understand a very small portion of the truth. With our senses and our consciousness, we only accept the present truth. However, we now know that the past and future exist right now, in this very second. Even the term *second* is a construct, since there is no time, and so there really are no seconds.

The linear course of time is an idea that we have invented.

- Time is not linear. Everything occurs simultaneously.
- Therefore, since everything occurs simultaneously, our future can affect our past.
- There is not only one reality, there are countless realities.

As there is not only one reality but many different ones occurring simultaneously, we can be influenced accordingly by many echo waves from the future. With that, many different possibilities are available that open our future to development at the same time as our present and our past. Strictly speaking, our present develops from the offer waves that we sent out in our past. This can be imagined as if we were in a movie whose ending depended on our decisions. In other words, we have the "happy ending" in our hands.

Every possible future sends its signals back, and these meet our offer waves.

Not all of these waves go together, and not all of them resonate with one another. This modulation of offer wave and echo waves occurs only when these wave forms are similar to one another. Only then does a so-called transition occur, which is the connection between our past, i.e., that which we sent out, and one of the many possibilities of the future. Only then a very high probability of an event is produced.

Where do the waves from the future come from?
    Physicist Fred Alan Wolf reached the following conclusion: every person who consciously or subconsciously sends out energy, which includes everything he or she holds to be true or thinks or is convinced of, sends a offer wave into the future in the manner already discussed, *but simultaneously also sends it into the past.*
    Strictly speaking, this wave works spherically, and not only spatially but also chronologically. At the same time, echo waves are also sent from the future. Future events also send echo waves backward, into our present. Thus the future exists in countless versions "in which countless variations of possibilities exist and send echo waves back into their past—my present."*

The offer waves transmitted by us search for equally oscillating echo waves from our future in order to build a high possibility of event probability.

---

*Jörg Starkmuth, *The Making of Reality.*

The energy we send out tests all possibilities of our different futures and creates a connection with the wave form of the future that is most like our offer wave.

Our offer wave pervades our entire future. All possibilities are tested, both in the next blink of an eye, as well as in the next year or the next century.

Quantum physics has discovered the following phenomenon: the chronologically closer a future event is, the higher is its resonance clarity. Put another way, "The chronologically closer an observed event is with the future, the clearer the decision if an event will happen or not."*

Our current level of consciousness determines all possible events in the near future.

If the anticipated events occur in a more distant future, resonance clarity is diminished; meanwhile, our current level of consciousness presents a clear direction. Since there are countless possibilities regarding levels of consciousness that send out energies, we can connect with each possibility. We do this as our offer waves communicate with echo waves and create an event possibility. In this way we are directly reached by our wishes. And so wishing is nothing other than summoning from the many possibilities in our life.

- When we wish, or create a very precise vision, we send out offer waves.
- They communicate with an echo wave.
- When we achieve an event possibility, we have the best chances of fulfilling our wishes.

---

*Jörg Starkmuth, *The Making of Reality*.

Everything we possess in our inner world
Will also be encountered in the outer world,
Because the outer world mirrors
Our inner consciousness.

Only when we align our
Consciousness in a targeted way
Can we step into resonance with things
That we would like
To realize in our life.

If we want to get results
That we wish for
We should begin
To observe and control
Our thoughts, feelings, and convictions,
Because everything we think and feel,
EVERYTHING,
Evokes a resonance field.

Can Matter Also Be Changed by the Power of Thought?

We are what we think. All that we are arises with our thoughts. With our thoughts we make the world.

BUDDHA

In previous chapters we learned about how strong and powerful the energy sent out by the heart is. When it comes to our resonance field neither time nor distance play a role in getting to where a similar oscillating field is present. We also learned about the power of thoughts to influence the DNA of others.

These discoveries are astounding and they have a much broader impact on our lives and on our *successful wishing energy* than we may have thought, because the strong energy sent out by the heart possesses enough power to not only change one's own body but also change the atoms of physical matter that surround us.

To better understand this, let's take a little detour into the language of atoms. Whoever has not visited a classroom in a long time or read a book about the new physics probably still has the old mechanical model in their head, i.e., there is a nucleus, which is surrounded by protons and electrons. This model contains solid particles.

But this model of matter has been obsolete for a long time. The mechanical image of particles circling one another, looking like a small solar system, has been replaced by a new paradigm, one based on new discoveries. We now know that our microcosmic world does not involve small discrete particles, but rather a concentration of energies in different zones. The closer we get to electrons and protons and nucleii, the more they are researched, the clearer it becomes that the thickest, strongest "particle" is really only made up of energy swirls. In other words, *the atom does not have a physical structure.* It is rather a swirling circle of radiating energy. These small energy swirls are known as *quarks* or *photons*.

There is no permanent matter; there is only energy.

This energy consists of electrical and magnetic fields.
Let us now return to a discussion of the energy transmitted by the heart. It is in fact in all cases the same energy field: the energy of atoms consists of the force fields that we generate with our thoughts, feelings, and convictions of the heart.

Atoms are changeable.

In 1896, Pieter Zeeman, who in 1902 won the Nobel Prize in physics, discovered that a *magnetic* force can change the fabric of matter. A few years later, in 1913, Johannes Stark, another Nobel Prize winner in physics, discovered that an *electrical* field can also affect an atom. It could then be concluded that an electrical field has the same transformative effects on an atom as a magnetic field.

From countless experiments in quantum physics we now know that the atom itself changes when the electrical and/or magnetic field of an atom changes. This means that atoms change their behavior and thereby change their expression as matter.

As previously mentioned, the heart energy possesses enough strength to influence atoms. This means that we can actually change every atom in the world. And neither time nor distance can hinder this process, as long as the heart's energy is directing the change.

Naturally, this is a fantastic concept! From now on we don't have to rely solely on our faith, because it is scientifically proven that we have a direct impact on our environment.

Perhaps we can now begin to grasp what powerful keys for our future we hold in our hands. Knowing this, the most unbelievable things can occur. In our mind they may still be inexplicable miracles, but for science and the spiritual leaders of the world there has long been no more mystery.

Dear Mr. Franckh,
Something impossible happened today!
　For several weeks, we have been planning to purchase a house. To do this, we had to sell our shares of some stocks. Yesterday, we had a shock: a stock crash. All the stocks hit bottom. I could not sleep that night; I kept thinking that we would no longer be able to afford our house. And then I wished that our stocks would rise before we sold them off. I held that thought the entire night.
　The next day, I received a phone call from a man saying that Volkswagen stocks rose for inexplicable reasons. Within an hour, we were able to sell our stocks and pocket a gain of 5,000 Euros. That same evening, the stock prices clearly fell again.

Greetings—and thanks!
Petra

*Find out
What your innermost convictions are,
And you have the key
To your life in your hands.*

Connected to Everything

The happiness of your life depends upon the quality of your thoughts.
　　　　　　　　　　　　Marcus Aurelius

Whether through our heart energy, our DNA, or our brain, we constantly send impulses to the outer world through our thoughts, consciously or unconsciously, which then meet with similar energies being sent by other people. Their energies cannot do anything other than oscillate with our energy, all of which is in the same resonance field. Otherwise, if others are not resonating the same as we are, they cannot be touched by our energy. Naturally, we are simultaneously drawn to their energies when they oscillate with our own. In this way we are continuously senders and receivers.

Through this connection with all others, we can shuffle the cards of our lives, starting from scratch. It gives us the opportunity to change everything in our lives—whether we are looking for a deep, true love relationship, a cure to an illness, or certain material things in our life. Because we are connected with everything, our resonance field brings us, through our thoughts and beliefs, everything we want, and even allows us to affect the resonance fields of others.

If we apply this ability specifically, we have the possibility of changing our life according to our will. But to do this it is a prerequisite that we know the true contents of our convictions and thoughts, and therefore learn to direct them consciously so that doubts or feelings of inferiority cannot find us.

Our convictions are always with us.

Everything we believe will manifest in our life, regardless of whether it involves the healing of an illness or a completely unusual or unforeseen success. This happens whether it involves our anxieties or our deepest desires. Everything of which we are convinced will occur. This means that our fears will come true if we send our energy to them.

The source of our reality can be found in the casual power of our thoughts and the law of resonance, which draws into our life all the hopes and fears that oscillate around us.

It will happen according to how you wish it.

The more we believe,
That we are separated from everything around us,
That we are powerless to take a stance against things,
That we are a victim,
That everything is random and that our life is an
    unmanageable chaotic sequence,
That the world is unjust,
That we have no influence on our illnesses,
That we are not due anything in our life,
That luck or misfortune is randomly distributed,
That our body is foreign to us,
That we have no influence on our own creativity,
The more we move exactly into this frightening life.

At the same time, we will feel an emptiness and loneliness within, because we have distanced ourselves from the creativity that stirs within each and every one of us, and with that we will distance ourselves from our own divinity.

It is no surprise that out of this loneliness we try to fill our life with superficial things. We set goals that could never make us happy because they only touch the surface. As long as we continue to be separated from our own creativity, as long as we don't see ourselves as the creators of our own universe, our life will seem to be a flow of random events. But even in such a case it is *we* who are drawing these "random" events into our life. We ourselves create the conditions about which we complain so much. This initial potential lies within us. Nothing in this world can exist without this original energy.

The law of resonance always says "yes." It is confirmed by our beliefs.

Energy does not ask about morals or how we will use it. Energy only reacts to our transmitted impulses.

Now what are the consequences? The message is clear, and hopeful:

At any time we can get out of the world we have created and create a new world.

We do not need to do anything other than change our perspective a little bit. We can begin to do this by starting to see our daily life with different eyes.

The key lies in understanding the ways in which we are connected with everything around us and in understanding how we can consciously change our resonance field to draw into our life the experiences we want.

Once we see ourselves as being a part of this world, part of a whole, and not anymore as a separate, discrete entity, we have taken the first and most important step in drawing all our wishes and desires into our life. From then on, we will no longer indulge in any more careless negative thoughts, because we know that these thoughts generate a resonance field that connects with others with similar thoughts, thereby influencing our life in a way we do *not* want.

When we understand the ways in which we are connected to everything, we have access to the greatest power of the universe.

When we change our patterns of belief, our old, created convictions and opinions about ourselves, then we take those people and events that have been oscillating in this negative energy field out of oscillation, which means that they will disappear from our life. Instead, we bring into our environment new people and experiences that are brought into oscillation with us through our positive thoughts.

Reversing one's thought patterns is described most impressively in the story of Maria, who found herself in litigation for eight and a half years, until one day . . .

> I was searching for a gift for a friend. Then I bought your book Wünsch es dir einfach—aber richtig (Simply Wish It—But Correctly). I immediately forgot everything else and walked out of the bookstore with an overjoyed grin on my face. I'd never known luck, but I felt I had finally found a solution for my problem!
>
> My problem? Well, I had the audacity to sue a lawyer. It took no less than eight and a half years! Yes, I won the first hearing, but he did not want to believe it. So he did not want to pay.
>
> I did not give up and insisted on continuing and sued again.
>
> The Court of Appeals took my case. With that came a seemingly never-ending trembling game. I clammed up from the anxiety I was experiencing.
>
> And then this high-spirited wishing book just smiled at me . . .
>
> I immediately followed all the "rules." Luckily, I did not have any problems believing in the power of positive wishing.
>
> On the day of the court date, I sat in my car with an unprecedented calm. I had driven to court singing my affirmations the entire way and was simply happy.

My lawyer was completely surprised, because she had seen me completely stressed out. I told myself that we would win and that she would only look at me speechlessly.

Then everything happened very quickly. Within forty minutes, the judgment came through, and we had good reason to pop the champagne cork: I received damages of approximately 140,000 euros. And this time I got it to the cent! Hooray! The opposing party did not have any further objections.

Thank you from the bottom of my heart for all the tips in your book!

Maria

Prior to her court date Maria became convinced of her success and sent this energy out without an ounce of doubt. Since beliefs sent out are 5,000 times stronger, it was clear to her that her success *would* occur. And it did, as she remained flowing, easy, and relaxed.

Before we begin wishing,
First comes our decision
As to which direction our life should take.
Any direction is possible.
Any.

Think about who you would like to be,
Then determine who you are.

If there is a big difference
Between these two people,
You should quickly direct
Your beliefs to the person
Who you want to be,
And no longer identify with the person
You certainly do not want to be.

PART 2

Effective Ways to Come into Resonance with Your Wishes

What Resonance Field Do You Exist In?

Pay attention to your thoughts, they are the beginning of your actions.

CHINESE SAYING

In part 2 of this book you will find various practices and examples that show you how you can create the correct energy field for successful wishing in the fastest way possible. Be aware that we are always dealing with generating the *correct* resonance field in order to send out our wishing energy. When you accomplish this, you can do anything in this world.

To come into resonance with our desires, we have many options. In the following chapters I introduce some of the techniques that have been easy and effective for me in the past.

It is best that you use those manifestation techniques that make you the happiest and that you can readily believe in. Ease and joy are the crucial elements around which our desires coalesce and manifest. If we approach any of these techniques seriously or with tension, we will inevitably create a negative, tense resonance field. What we will then draw into our life will follow in kind: seriousness and tension. So keep it light.

Ease and joy are the crucial elements around which our desires coalesce and manifest.

Not all the methods I include in this book will succeed right away or become your favorites. Maybe with some of the techniques you might feel some inner resistance or sadness or doubt. Or maybe you simply cannot generate a belief in an affirmation, or cannot conjure up strong inner images. Do not be disappointed if this happens, and certainly do not let it anger or frustrate you, because this, in following the law of resonance, will only strengthen anger or frustration in you. Always, always remember:

Everything you think, feel, and believe, you draw into your life.

It is not necessary to jump from one method to the next or to use all of them simultaneously. Once you discover one technique that works for you and you have full confidence in that technique, you can accomplish more than if you were to use many different techniques half-heartedly.

The more fun and joy you have with your manifestation technique, the faster you will realize the results.

Before we begin with the wishing techniques, we first need to take a little inventory, by checking in with ourselves to discover which resonance field we are already in. This is relatively easy to find out, and I have included an exercise in the next chapter that will assist you in doing so. Since things and people found in our outer world are always a reflection of what's in our inner world, our immediate environment is a very good place to determine which resonance field we have created up till now.

Why would we even want to take such an inventory?

If we are surrounding ourselves with the wrong resonance field—i.e., wrong for us—we can make a concerted effort, over and over again, to rise up toward our optimal desired resonance, but our surroundings will always bring us back to where we essentially do not want to be, thereby defeating our efforts to cultivate a new resonance field. And everything in our life would continue to remind us that we are single, lonely, chaotic, poor, a loser, a mama's boy, incompetent, etc.

The resonance field that surrounds us influences our personal resonance field.

*Our resonance is a kind of matrix
That exists deep within us.
It sends information outward
That corresponds with our essence
And draws only the similar into our life.*

*We can therefore use our environment
As a type of "reading system"
To understand our resonanace.*

*From the one who comes to us
Or the one whom we draw in,
We can read how we are programmed
In our innermost being.*

*Our connection with all things
Helps us understand ourselves better
And fulfill our potential.*

When Resonance Fields Hinder Your Further Development

There is only one time when it is essential to wake up. That time is now.

BUDDHA

We know that we can be influenced by resonance fields of others. We all know this too well in our daily lives.

If, for example, we are peaceful and calm, it only takes one single angry person to carry us along with that person's anger and unhappiness. Within seconds, we may suddenly find ourselves in an unbelievably intense argument. It can erupt like a brush fire. We might say things that we do not want to say or make rash decisions that we would never make under prudent consideration. Often we do not understand how we allowed ourselves to get into such a heated argument, since our day was wonderful and harmonious and we were feeling great up until that point.

Yet the explanation is quite simple: we were unconsciously pulled into the resonance field of that person and therefore influenced by him or her. So, despite our earlier peaceful state, our oscillation energy was in tune with someone else's. If we had not been in tune with that person's energy we would not have been influenced by it.

And so, if we look at things accurately, we have to conclude that this energy was actually not foreign, not outside of us. If it had been, it could not have influenced us. We would not resonate with it. The allegedly *foreign* oscillation would not oscillate with anything in us.

If we let ourselves be infected by quarrelsome people, the potential for a fight already lies within us.

We can be anything we want to be. All facets of human emotion lie within us. We can be friendly, loving, sensitive, angry, sullen, jealous, moody, doubting, etc. When we encounter someone full of love and compassion, then we too can become soft and full of love and feel this same compassion within us. Both anger and compassion are already in us, but we cannot come into both of these energies at the same time, as these are opposing energies.

It lies within us which inner oscillation potential we want to activate.

It's all up to us.

There are many examples of how people allow themselves to be influenced by foreign energies. For instance, when we enter a church or a deeply spiritual place, we immediately feel different, not only because this is expected of us, but because we have entered the calm and peaceful resonance of that space. We "catch" this resonance and are immediately calmer and more peaceful. Sometimes, our inner stillness can remain turned on even when we go back out on the street and return to our daily life.

Something similar happens when we read a spiritual or inspirational book or hear uplifting music. We immediately discover the resonance of the book or the music and align ourselves with it.

We simply make use of the existing resonance field that is nearer to our desired energy.

We should stop and take note when we find ourselves in a resonance field that goes completely against our wishes. If we are surrounded by people who weaken us with their doubts and want to convince us that the things we want will never work out, it will be no surprise if we begin to doubt ourselves, as we are allowing ourselves to be drawn into the resonance field of doubters. This development can often be subtle and insidious. It is therefore good to occasionally double-check to see if we are in the correct, beneficial resonance field.

At one of my lectures I met a woman who told me she had written a book and wanted to be coached on how to find a suitable publisher. She told me she had hired a woman who was a very successful author to help her to achieve this. The woman did not seem to be excited or happy that she had just written a book. Instead, she was worried because her coach had given her the understanding from the outset that she should in no way imagine that her very first book would quickly climb to the bestseller list. This was highly unlikely, she'd been told. Ultimately, her coach said, even she, a successful author, had published three books before she made it onto the bestseller list.

This woman wanted to hear my opinion as to whether she should let her coach continue to help her or not. I asked her if she felt encouraged by this woman. My next question was who was paying whom? Naturally, she was paying her coach. I persisted: "Why do you pay someone who does not encourage you? Why don't you search for someone who sends you encouragement? Someone who gives you enough strength and euphoria to pass through all the obstacles and difficulties on the way to publication? It's like watching a terrible show on television, one we do not like at all, yet continue to watch only because we paid our cable fee.

When we realize that someone is not good for us, no one can force us to stay with that person and be around their negative energy.

We might think that we will not find anyone who will support us in our dreams; this, however, whispers to us only from the mind that is not responsible for miracles. Our mind usually confirms our self-doubts and feelings of inferiority. So if we truly want to come out of that, then we quickly need to find a new, positive resonance field, one that allows us to become bigger and to grow into our full potential.

Surround yourself with people who motivate you!

Surround yourself only with people who believe in you, who are convinced of your strength, who see that wonderful talent in you, who recognize your potential, who connect with your visions and further these visions.

Why should you stay with people who halt your enthusiasm and do not believe in your creativity? Who has told you that you have to endure that? Our lives should be fluid, easy, and serene, carried along by positive, powerful energy, by energy that advances us, that helps us grow so that we can outgrow our limitations and can dare to take new steps. But unfortunately we often surround ourselves with people who block us in our own development, because this is what we've grown accustomed to; it's what we're familiar with. Maybe we know this kind of energy from our parents, our siblings, our relatives and acquaintances. Such an energy, even if it impacts us negatively, somehow still feels like being "at home" to us, so we stick with it.

If we continue to adhere to this negative resonance field, then we will have a very hard time creating a new resonance field, one that enables us to draw positive experiences into our lives.

When I spoke to the woman at my lecture about this, she laughed and recognized that she had taken a turn onto a dead-end street. She suddenly realized she had given the woman who was her coach power over her, and she had even paid this woman to do so.

Maybe you've had a similar experience. We do not always recognize that a negative resonance field surrounds us and only see this if we examine our surroundings consciously.

There is a very simple, quick way to get clarity about the energies that surround you:

- Write down the names of all your friends, acquaintances, and relatives.
- Then note next to their name the characteristics you pick up from these people. For example, "supportive," "funny," "cheerful," "inspiring," "constantly critical," "judgmental," "jealous," etc.
- Don't overthink it. Write the first thing that comes to your mind. The first thing is always the right thing. If you think about it too long, it could be that your mind is trying to meddle and qualify everything.

In this way, you can very quickly determine if the people who surround you truly give you a feeling of strength or if they limit you and infect you with feelings of inadequacy. In case the picture that emerges is not easy for you to swallow, follow up with these questions:

- Is the predominant energy in your surroundings beneficial for your intentions or really destructive?
- Do your surroundings send you strength and energy?
- Do you feel well and secure there?
- Do people trust you?
- Do people support you?
- Can you express your wishes?
- Do people think about your well-being?
- Can you show yourself as you truly are?

It should come as no surprise if the answers to these questions are rather sobering. If it were different, you would already be in another resonance field, the one you want to be in, and you would be drawing the corresponding affirming energies and people into your life.

Next, let's go a little further in our exercise:

- Cross off the list the people who do *not* believe in your power and creativity.
- Perhaps you would like to think about how much longer you are willing to allow other people to judge you, devalue you, or influence you negatively. Write down a number, an exact time: date, month, year, or for the rest of your life. With that, you can recognize well what your real projects in your life are and if you are furthering them yourself.

There is only one person in this entire world who gives you permission to allow this, and *you* are that person. Only *you* invite other people to have the pleasure of working in your life. And it is always *your* decision to eliminate the negative influences in your life.

Let's go back to our exercise:

- Again, list all the people who support you, protect you, and are always on your side. Even if there are only a few such people, there is always *someone* on your side, someone who encourages you unconditionally. Think about it. Maybe you have forgotten this friend or you placed a different emphasis in your life that made you overlook this person. Maybe you understand this friendship as being obvious.
- Decide to devote more priority and time to the people who are beneficial to you and who continuously bring something good into your life.

Surround yourself with people who are sympathetic toward you and who send you appreciation, respect, and support.

- If you do not have such people in your life yet, then use your mental resonance field to attract such people.
- Be aware that people who recognize your full potential exist.

- Be sure that these people enter into your life now.
- And most important, recognize your own potential.
- Begin to be aware and love and treat yourself with full appreciation. The more you do this with purpose, the quicker you will change your environment.
- The greatest aid to entering a desired resonance field is to encourage others in their own dreams. The more one gives, the more one gets back. Because actions always draw in similar actions, you will quickly be surrounded by people who mirror your inner generosity and also want to help advance you.

Embark into the field of resonance you desire.

Naturally, this does not mean that you must immediately abandon your family in order to realize your wishes. It does means that you need *to add* to your life those people who will support you in your wishes. Otherwise, you may increase your resonance energy through affirmations or strengthen your visionary powers through targeted body exercises, but the resonance field you generate will always be somewhat disturbed or will feel stifled or even completely fall apart if you continue to surround yourself with people who do not believe in the power and potential of your wishes.

Negative resonance fields can disturb our development.

Conversely, positive resonance fields will positively affect us and bring something positive within us into oscillation, which is very important for our development. The choice is clear: we should use such positive resonance fields to bring our wishes into fruition.

The world is open to you. The world has always been open to you. You have to open the door just a crack, so the energy can flow to you. Once you are open and ready, the resonance field will do the rest.

The Gift of Mirror Neurons

Our wishes are the harbingers of the abilities that lie within us.

JOHANN WOLFGANG VON GOETHE

In 1990, researchers under the direction of neurophysiologist Giacomo Rizzolatti, a professor of human physiology at the University of Parma, Italy, and the author of many books in the field of cognitive science, discovered a fascinating phenomenon. They found that specific regions of the brain fire both when the animal acts and when that animal observes the same action performed by another animal. Thus, these "mirror neurons," as they have since been dubbed, mirror the behavior of the other, as though the observer itself were acting.

Our brain records memories of specific movements even when we never experienced these things ourselves.

This was completely new and surprising for science. These recorded memories—recorded purely as a result of observation—enable us to actually perform similar movements, although we may have never actually physically done them. Although the experience of performing these new skills is completely missing from our reality, we somehow know how to behave.

These mirror neurons are activated when we want to execute very specific actions, such as balancing on a rope or something else unusual.

They are activated—and this is interesting—when we watch how someone balances on a rope or does something else unusual. Mirror neurons award us with the ability to inwardly understand actions by simply observing others.

Medical science now holds this to be very successful. For example, stroke patients can learn from movies that show a person's daily hand motions (turning a faucet on, etc.) how to go about making these motions themselves.

As a youth, I had an experience that today I attribute to mirror neurons. At the time I was occupied with playing the guitar, and I was more bad than good. I had a guitar teacher who unfortunately did not inspire me, and therefore I made no real advancements. But then I attended a concert one night by Manitas de Plata, the French flamenco guitar genius.

I was thrilled by de Plata's performance. I sat in the first row, completely fascinated, and absorbed every single note. Later that same night, I played those same Spanish guitar riffs and was better than anyone else. It was as if I had already practiced all the techniques. I was fascinated by my newfound ability for Spanish music, seemingly acquired overnight, and also bewildered by this sudden rush of knowledge. Now I know where it came from.

- When we want to improve our own skills, we only need to observe others intensively as they perform these tasks in order to incorporate this knowledge into our own experience.
- Mirror neurons help us make exceptionally fast improvements.
- We should take advantage of mirror neurons to accomplish our goals. However, these neurons are only the first steps toward our goals.

In observing, we sometimes transcend time and space differences and we can feel we are actually present with another, body and spirit.

The same mirror neurons are responsible for this. For example, when we identify with an athlete, we might begin to cheer him on. Without consciously realizing it, we are taking in those physical actions as if we ourselves were performing them. If the situation for our favorite athlete is critical or the competition reaches a decisive phase—for example, during a boxing match, a fencing tournament, or a penalty shootout in soccer—we too tense our muscles, our pulse accelerates, and we breathe faster. Sitting at home, far away from the stadium where the athlete is performing, we become excited—we yell, rave, laugh, suffer, and rejoice. We experience everything as if we ourselves are standing there in the arena, yet we are only sitting in front of the television at home, with a bag of chips in hand. Our entire body is involved.

Mirror neurons give us an opportunity to empathize with other people. They are responsible for eliminating the feeling of separation we may have between ourselves and others. We feel the feelings of the other—compassion, disgust, joy, etc.—within ourselves and can therefore understand these feelings in others. Most of all, as we see others modeling an experience, it is as if we ourselves are experiencing the same thing.

If we fully act on and take in the feelings that we observe, they will later be at our disposal as memories.

The activity of mirror neurons tells only part of the story as to how we can realize our wishes. The other part has to do with the power of visualization.

In the 1980s, Denis Waitley, an American motivational speaker, writer, and consultant on human productivity, undertook a visualization program in which he challenged participants in the Olympic Games to imagine the competition and precisely think it through in their own minds. To better evaluate the research, the participants were connected to biofeedback machines. The astounding results showed that although the subjects in this experiment only imagined these competitions, their muscles reacted as though their visualizations were reality, as if the competitions were actually occurring, including the same muscular activities involved.

This knowledge about mirror neurons is something many athletes have since adopted into their training regimens. It's called mental training. Long before the actual competition takes place, the athletes go through all the important motions in their head until they become second nature. They visualize and recap certain experiences, but only in their thoughts. Thanks to mirror neurons, the sequence of events is at their disposal in a matter of seconds. It is then as if the athlete had already performed in the competition, as if she or he had already done it a thousand times and had much more experience, responsiveness, and foresight.

For this reason you see athletes make an entire competition run in their mind's eye. For example, the 100-meter runner at the starting line has already imagined the entire sequence of events. The runner's body has played everything through mentally, and has thus prepared every cell in the body for the event. Most importantly, the runner can recall 100 percent of the exact sequence of physical movements he or she visualized and put them into action when the starting gun goes off.

If you ask a sprinter how he achieved his goal, how he was motivated for such a long period of time and how he might have found power that others might not have believed he had, he, like many athletes, might say that he only concentrated on the goal, what he wanted to achieve. The joy of achieving that goal was so intense that all the steps to get there seemed small and relatively insignificant. The athlete was intimately connected with the goal and with the joy of winning.

We need to build the resonance field of winning.

The more intensely we imagine our goal, the more intensely we think about "winning," the more we are able to see it in our mind's eye, the more we will be able to develop the resonance field that will attract that goal. Within that resonance field we attract all those in our lives that help us achieve our goal. In addition, our entire body is already resonating in the energy of the desired event.

We actually know today that we can create a kind of library of neurological operations in our brain when we use our full imaginary powers in this way. From this perspective, a deeper understanding of how visualization works is possible. When we consciously, with the help of our imaginary powers, place ourselves in scenarios in which we achieve a desired outcome, we optimize our resonance field and strengthen our convictions. In so doing we activate our mirror neurons. In terms of our brain, it is the same as if we already had sufficient experience with such situations; it knows exactly when to do the important things and impresses on us and on our environment a feeling of confidence, calm, and knowing. Consequently, our body feels secure in our inner convictions.

I personally made it a habit a long time ago to imagine important meetings or appointments and imagine the desired outcome beforehand. In this way, I always feel safe and confident and always know how to respond quickly or react with ease in any situation and thereby help shape the outcome.

Mirror neurons send us a type of memory, as if we have already sensed these things ourselves. Maybe you remember the feelings you experienced thinking of other people who came from nothing to a position of wealth or prestige. Maybe at the time you felt inspired and motivated and felt that you too could do that. Yes, mirror neurons are also responsible for these kinds of experiences. They record the experiences of another person you hear or read about, and then you see and feel those experiences to be your own. It has been found that the more one identifies with someone else and the more similar to that person your desires are, the more intense the recorded experience is.

This whole concept introduces us to possibilities that we never thought we had at our disposal. That is why role models such as Gandhi, Buddha,

Jesus, Martin Luther King Jr., Mother Teresa, Nelson Mandela, and other pioneering personalities are so important for our development, because they inspire our consciousness and guide us in a new direction. The same thing occurs when we read about or hear the success stories of other people—our mirror neurons help us internalize what we read or hear until what that person achieved becomes a reality for us, too. Success stories therefore help us out of the boundaries we may have set for ourselves. If someone else did it, we know that we too can achieve the same thing.

Our mirror neurons become active when we mentally deal with things that seem physically impossible to others.

Whatever you would like to achieve, bring in the experience of how others achieved it. Read it, see it in yourself, analyze it, and most important, make their experience your own.

I once read in a study that the favorite genre of literature of extremely successful people is biography, specifically, the biographies of successful people. With the knowledge of how mirror neurons work, this isn't surprising. The more we read about successful people, the more we record our *own* memories in the brain. We suddenly know that everything is possible. There are no limitations. *Obstacles are only hurdles for even greater leaps.*

I encourage readers to explore the stories of those they admire.

- Occupy yourself with the lives of successful people.
- Read biographies or see movies that describe such ways of life.
- Explore the stories of people who went from financial straits to financial freedom.
- Occupy yourself with wonders.

It is not enough to have knowledge,
One must also apply it.
It is not enough to have wishes,
One must also accomplish.

JOHANN WOLFGANG VON GOETHE

Surrender to the Field of Resonance

Do not spend your time looking for an obstacle, perhaps there isn't one.

FRANZ KAFKA

Be prepared to have physical and emotional contact with your wishes. To use the gift of mirror neurons optimally to achieve that which you desire, it is important to not only visualize your goals and wishes as being realized but to also physically put yourself in or near the resonance field of that person or event that corresponds to your wishing energy.

If you wish to be prosperous, surround yourself with people who have already been brought into prosperity. If you would like to have a happy, lasting partnership, try to be around couples who already live in harmony and satisfaction. If you would like to have a specific career—for example, if you want to be a doctor, technician, or whatever else—then surround yourself with people who are practitioners in your desired field. Not only will your own resonance frequency be moved toward the desired direction, you will also obtain new, heretofore unknown experiences as a result of the activity of your mirror neurons. In addition, you will, through your generated power of attraction, receive guidance and information as to the next steps you should take. You will receive clues that will help you continue. You will get to know people who will encourage you in the feasibility of your attaining your goals, and people who know that you can fulfill your dreams.

There is little that is more motivating than the success of others. We too can achieve what others have achieved! Most important, those who have already achieved these things can understand our desires—after all, they themselves had them at one point. And they will support us and—very important—believe in our potential. Whatever you want to realize in your life, go to where it has already been realized for others.

Surround yourself with people who believe in you. You will find them easiest where the fulfillment of your wishes lies.

For example, if you would like to get an apartment in a better building, then stop in the neighborhood you like more frequently. Go walking there, visit the cafés and stores there, wander through the park there, enjoy the atmosphere and feel the energy that prevails in the community. Let yourself be captured and infected by these new, different oscillations. Imagine how good it would be to live there. Visit apartments in which would you like to live, even if you cannot rent them yet. And when you're there, go out onto the balcony and rejoice in the great view. This energy is available to you!

Create a sense of anticipation, such that you not only can see what it is you desire, you can physically feel it. This will raise the vibration in your resonance field until what it is you want becomes all-encompassing, thoroughly pervading your resonance field.

- The more time you spend where you eventually want to be, the faster you will find yourself in that role or that place.
- If you remain in the place that you want to leave, sitting there anxiously, you will always be fighting the oscillations that surround you. You must be stronger and more powerful than those oscillations.
- If your home environment does not represent what you want in your life, go out on the streets, to where the desired energy can be found.
- Let yourself be captivated with a new, different oscillation. It will then be easier to make changes in your own life because you already brought yourself into connection with anticipation.

- By immersing yourself in your desired resonance field, you will have opportunities to access information that will reveal your next steps. Maybe you will meet people who invite you into their realm, give you advice, or even offer you their help.

When you cannot get out of the cycle of fear and worry, then at least for a short while simply leave the physical location that is keeping you in that oscillation.

Sometimes people need only a few hours' break to renew their energy. Maybe you allow yourself a full week break. We have all experienced how quickly our perspective can change with a change of scenery. For example, after a vacation we often come back full of new ideas, visions, and realizations that we collected while we were away, and that we would now like to unpack. In the beginning we have a lot of courage and zest, which can sometimes become paralyzed within a very short amount of time.

While on vacation, we did nothing else other than change our resonance field, even if doing so was subconscious. But after a short amount of time, back in the old location, we find ourselves getting caught up in the old, familiar oscillations of home. Every now and then, we dream of the vacation and about the projects that have not yet been achieved. Sometimes we can even get depressed about it, feeling we do not possess perseverance.

Instead of letting ourselves become discouraged, we should be motivated with the knowledge that it is easy to change our resonance field when we are in the right location.

- Consider which of your wishes you would like to realize next and which places you should seek out to generate the correct, corresponding oscillations, the ones you need to realize your wish.

It is not important how crazy this might make you feel. If you would like to be married, try on wedding dresses; test drive the car that you have your heart set on, even if you cannot afford it right now; visit apartments in neighborhoods where you would like to live; get into extensive discussions at the travel agency if you want to travel. Internet forums are also good places to find the energy from others that supports you in your wishing. Don't just keep your wishes inside you; let them lead you to an actual physical experience.

Make physical and emotional contact with your wishes.

> *We do not create our destiny; we participate in its unfolding. Synchronicity works as a catalyst toward the working out of that destiny.*
>
> — DAVID RICHO,
> *THE POWER OF COINCIDENCE:*
> *HOW LIFE SHOWS US WHAT WE NEED TO KNOW*

*Every form or structure,
Whether outside in nature
Or as a part of our bodily functions,
Vibrates and pulses in an
interdependent relationship.
This vibration field is called resonance.*

How Does Your Resonance Field React to News Media?

> *Every piece of information, every communication, every message has an influence on our DNA and leaves an impression in our collective cell structure.*
>
> PIERRE FRANCKH

The news, whether on TV or in a newspaper, is nothing but a collection of negative communications. We learn about unemployment, floods, climate catastrophes, energy crises, bank closures, terrorist attacks, and the possibility that we ourselves could be the target of such an attack. This is only a small selection of the terrible scenarios with which we are sent to bed night after night. After we hear the worst things imaginable, accompanied by the cruelest images that could be gathered from around the world, the news anchor smiles and wishes us a good night.

What would a *good night* look like?

TV news and newspapers are the *best* transmitters of fear, stirring up our collective anxiety and strengthening our disillusionment. Through them we remain in a perpetual state of anxiety and self-protective fear, which our body can become overwhelmed by. Reading the newspaper or watching the evening news, we are guaranteed to always feel weaker and more powerless, no longer seeing things with any kind of perspective.

Remember, like attracts like. When we get caught up in fear energy, we draw in the experiences that confirm our fears.

Through our fears, we create exactly that which scares us.

The news has an insidious effect. It is not at all easy to get out of the cycle of fear and low- or high-level anxiety induced by the media because the activation of stress hormones shrinks our ability to think clearly. Furthermore, every information byte, every communication, every message, has an influence on our DNA and leaves an impression on our cell structure.

All information storage and problem-solving takes place in the forebrain. This is the seat of reason and logic. Reflexive, logical activity has its place in the hindbrain. With that, stress hormones can go to work very quickly in an emergency, constricting the blood vessels in the forebrain. The result is we don't think clearly anymore. Beyond that, stress suppresses the center of deliberate action, which is in the front cerebral cortex.

Under stress, we actually have reduced intelligence and conscious perception.

Under stress, we cannot think clearly and make conscious decisions. It is therefore extremely hard to come out of fear energy and generate a positive resonance field.

But that's not all. When we fall asleep with fear and stress, we take this energy into our sleep awareness and store it. Current research on the brain has revealed that what goes into our consciousness is stored in our memory *at night*. This means that the closer the experience occurs to when you go to sleep, the more intense it will be built into your memory, and when the experience involves fear and anxiety, it will be that much closer to your reality. That which might seem harmless really has an enormous significance.

How does one get out of this cycle? Very easily—simply by exiting.

- Try to see what it feels like to go to bed without watching the news, and instead take some time to find your own positive thoughts. Do this for a whole week.

Have you ever noticed that politicians like to stir up fears right before a decisive election, so that they can then offer us promises that only *they* have the optimal solution, and we should therefore not worry? And let's not forget the news anchor who presents us with a dire picture of the world, and who then wishes us a good night.

Maybe *you* should try sending yourself into a good night. After all, your life lies in your own hands. And it belongs exactly there.

What good are all the wonderful resonance fields that you have created if you simply tear them down again?

Making a habit of the news is not the only thing that persistently interferes with our wishing energy. Let's take a look at our other viewing habits.

What kind of movies do you watch? Do you like crime movies, horror, political thrillers? Do you prefer dramas and tragedies? Does your movie hero have hopeless situations and illnesses, the death of beloved people, or financial ruin?

Always consider that the movie industry does everything it can to completely occupy your spirit. A truly good movie can only capture you when it seizes your emotions. Every director and screenwriter wants nothing else other than to touch the world of your emotions. Film producers therefore spend millions and employ an entire team so you will empathize with what their movie exemplifies. But remember, the mind cannot differentiate between fantasy and reality. The subconscious stores these experiences in its memory and directs future information based on these new experiences, whether they are fictional or not.

When we watch a movie, we are involved in drama—whether it's about war, famine, or a fight for survival—that may seem to last only ninety minutes, but the information goes into our memory storehouse. If we do this frequently, then we come to occupy ourselves with these things both mentally and emotionally. The more we watch these kinds of movies or watch the news or special broadcasts about bad luck or famine, the more we fortify a resonance field of poverty, worry, and fear. And this will work at odds with the positive resonance field we are trying to create.

When we:

- Continuously expose ourselves to TV shows in which tragedy is played out, people are killed, etc.
- Read books in which dramas are played out or people are murdered and we feel an inner tension and are captivated by the book
- Extensively and intensively absorb news or messages that shock us or that produce disgust or defense

we build this resonance field also within us. We build up an enormous resonance field of poverty, worry, helplessness, hatred, and fear in us. The law of attraction works and we identify more and more with what we do not want to be.

Sometimes—perhaps too often—the resulting feelings are much more intense and sustained than the feelings we have caused by our desires and wishes.

How different would your life be if for a week you only admitted energies that were conducive to your health and well-being? Try the following:

- Only read edifying literature.
- Watch only movies that give you courage and inspiration, or that make you laugh.
- Only listen to uplifting music.
- Only meet with people you really like.

- Write loving letters and e-mails.
- Write down your positive thoughts in a journal.

Let yourself be well.

Everything you expose yourself to influences your resonance field and awakens different emotions. When you hear pleasant music or read uplifting books or watch uplifting movies, your disposition becomes considerably friendlier, lighter, more cheerful, and calmer. These feelings pervade your resonance field. The synapses in our brain are responsible for this.

Expose yourself to those influences that demonstrate the fulfillment of your wishes.

The more often you do this, the faster those influences will become your reality.

*No one other than yourself is responsible
for the true wonders in your life.*

*Everything that we wish for,
All goals that we want to achieve
Already exist within us.*

*We can therefore
Step into resonance with our desired goals.*

*If it did not already exist within us,
The event, which is always there,
Could not be brought into resonance within us.*

The Brain Is Malleable

Imagination is more important than knowledge. For knowledge is limited to all we now know and understand, while imagination embraces the entire world and all there ever will be to know and understand.

ALBERT EINSTEIN

The latest discoveries of brain research are quite astonishing. We have now discovered that the brain is malleable. It changes not only theoretically but also physically. It changes after we pursue new thoughts or have new experiences.

In one study, test subjects were given an opportunity to experience unfamiliar activities for a certain amount of time. It was found that the area of the brain that was used for these activities became larger, similar to the way a muscle becomes larger when one undergoes physical training. At the same time, the other parts of the brain that were not being used in the new activities became smaller, with some parts of the brain actually shriveling. At the same time, new synapses connected, the flow of energy increased there, and new chains of thought began to be activated. The brain possesses the ability to completely change its networking system and create new neuron pathways when it does or thinks about something new for a specific amount of time.

In terms of the brain's function, it only takes a short amount of time for new activities, new thoughts, or new convictions to become powerful realities.

The brain realigns itself under different circumstances:

- When we think about specific things for a specific amount of time or do specific actions for a specific amount of time, the part of the brain responsible for this grows.
- Meanwhile, the portions of the brain that we set aside during this specific amount of time and do not use shrink.
- The neurons responsible for the new and necessary electrical signals for a new activity can change their function so that the information associated with the new activity can be transported more readily.

Science describes the surprising ability of the brain to completely change itself to accommodate new activities as *plasticity*.

The fascinating implications for this discovery mean that we can generate a new life reality at any given time. Our brain reacts to our thoughts and creates neurons afterward, reintegrating itself in a completely new way.

We can train our spirit, and with it, our entire future, in any direction.

When, over a specific period of time, we train previously fallow portions of our brain through new thoughts, our future experiences can proceed completely differently.

- We can activate new nerve cells in the brain through new thoughts and new actions.

- The functions of neurons can change and other logic operations can emerge, while regions of the brain that we do not use—for example, those that have to do with worry and other negative thoughts—become smaller and less important.
- This means we can completely change the course of our life.
- Even our convictions can change completely when we direct our thoughts in a new, desired direction.
- When our convictions change, our entire life changes.

Now we understand why it is so important to check our mental habits or repeated specific actions over time. To this end, affirmations, such as those that we will discuss in the next chapter, can help us reprogram our brain.

Note that the brain needs time to make changes. When the targeted nerve cells are stimulated, they build networks—in effect, small avenues—with their neighbor cells, and they do this within minutes. However, it takes a day for these avenues to become passable so that information can actually be exchanged. Neurobiologists at the Max Planck Institute have found that our nerve cells need up to twenty-four hours to exchange information over these newly constructed contact points.

Everything learned anew needs its own time. Within the first eight hours, small branches are built from the new networks that have been created. In the hours that follow, it will be determined whether these will need to be built up further or not. That is why if we want to retain new information, we need to reinforce the learning process by mentally repeating what we have just learned, i.e., "practice makes perfect." This is why we are told to repeat affirmations or positive convictions daily.

Only the consequent occupation with new, desired convictions allows us to neutralize old, undesired patterns.

Much to their surprise, neurologists have determined that the brain can actually forget old convictions. When a new neural pathway is created, it can replace an existing one—in effect, the possibility is quite high that this new pathway replaces an old one. From the new neural roots generated by your changing thoughts, you behave differently, reinforcing and strengthening the new neural pathways. The scientific term for this is *neural correlate of consciousness,* or NCC. An NCC is just a neural state that directly correlates with a conscious state, or which directly generates consciousness. It is generated by a certain way of thinking. It works like a loop: thought = root/pathway = action. "Neurons that fire together, wire together" for the purpose of creating a *"brain map."**

This explains why one should repeat a chosen affirmation until it becomes a conviction, to create a new, automatic thought chain. It might take a person one week to create a new way of thinking, or it might take longer; it all depends how strong the old thought pattern is. But the good news is you can change your life very easily and relatively fast, because, as neurologist Valentin Nägerl of the Max Planck Institute asserts, we forget old things when we learn something new. Furthermore, one can relearn things much faster than it takes to learn them the first time. This is an indication that old, suppressed logic operations are not completely wiped out, and that they can be reawakened when a situation demands it.

This knowledge clearly shows us that we can consciously change our concept of life at any given moment. It only requires a bit of time, some patience, and the repetition of desired objectives, which allows the brain to create new neural pathways.

So what new pathways would you like to create?

---

*Norman Doidge, *The Brain That Changes Itself: Stories of Personal Triumph from the Frontiers of Brain Science* (New York: Penguin, 2007).

All wish formulas and affirmations
Serve but one goal:
To program one's own resonance field.
It is therefore always about
Turning off old, negative belief patterns
And their effects on our life
And generating new, positive resonance fields.

The Power of Affirmations

Change yourself, and the world will change.
MALTESE PROVERB

Affirmations are positively formulated statements that one constantly repeats, like a mantra. They serve to strengthen and enhance one's life goals. But in reality, they affect significantly more than this. Affirmations are the fastest way to build the ideal resonance field to realize our wishes by reprogramming the brain, especially since they can be used anywhere and at any time.

Affirmations help us to transform our beliefs in the fastest way possible.

As we now know, affirmations that are thought or spoken over and over again wander deep into our subconscious and change our brain's default settings and our entire brain function. And that is the deeper meaning of affirmations. The mind begins to dissolve old programs and to organize new ones. We replace our hitherto sabotaging, negative beliefs with new, positive convictions.

Essentially, changes occur only when we say our affirmations with absolute conviction, because we draw that which we want into our life only when we *truly feel and believe*. So we must do more than make mental constructions when we repeat affirmations; we must *feel* the affirmation viscerally, all the way down to our skin and hair.

So to summarize, affirmations work in the following manner:

- Every affirmation is like a command sent to the subconscious.
- Affirmations change your old convictions.
- These new convictions are absorbed by the heart and the DNA, which in turn send them out.
- We communicate with all others in this world through our convictions. This occurs simultaneously in hyperspace, without any third-dimensional limitations of time or space.
- According to the law of resonance, everything that goes along with your new convictions will be drawn into your life.

Wishing statements—essentially, affirmations—not only help us to focus on our goal, they also affect our entire being. We change our beliefs in the direction of our wishes and send out this bundled energy into the world. The old saying "Faith moves mountains" takes on a whole new dimension in this way.

Which affirmations are the best?
This question cannot be answered generally. It's a matter of any one person's individual wishes, desires, and goals—and also that person's obstacles.

A good rule of thumb is that you should seek out those affirmations that *feel* best to you, those you have the least resistance to. An affirmation should send you a warm, comfortable, secure feeling. If when you say an affirmation you notice any inner resistance or find you do not believe what you're saying, change your wish sentence into something that you can think, believe, and say without any resistance.

Your affirmation should be easy for you to say and should send you a feeling of strength. If, however, you notice that it is strained and you must expend energy to say it, try going one step backward. For example, if your successful wish term is "I am rich" and you notice worries arising that can torpedo your wish, and you notice that you are not at all convinced that, for example, you are entitled to all the money you wish for, then change your affirmation. Maybe the sentence "Being rich feels good" or "I like money" would work better. As soon as you try a few affirmations, you will quickly discover where your blockages and negative thoughts are hidden. Use your awareness of those blocks to help you construct an affirmation that really works for you.

It is best to find affirmations that fit you well and possess power for you. You must feel the truth of your affirmation and it should convey a feeling of joy. For example, I have found that the statement "I am lovable" is unbelievably difficult for many people. Too many times people are convinced of the exact opposite, and this belief is continuously mirrored back to the person. If this is the case for you, before you become overwhelmed by the feelings that an affirmation like this might bring up—feelings that can throw you out of resonance and generate a contrary oscillation of self-doubt—try something you can actually accept and believe: "I like myself" or "I like myself more and more" can be quite powerful and something to which you'll feel no inner resistance.

How do I use affirmations?
Once you find the affirmations that work best for you, you can repeat them silently in your thoughts or recite them out loud. Feel the confidence that arises in you when you say the affirmation. You can say it like a mantra, over and over. Be sure to repeat it once you have found that you've generated a sufficient amount of feeling as you say it. If you notice that doubts creep into your thoughts, change your delivery of the wish statement and repeat it until you once again sense the power and freedom arising within you as you say the affirmation. In this way you can bring yourself back into resonance with the original wish and no longer leave room for doubts.

How can I empower my affirmations?
We can increase the strength of our transmitted energy in a big way if we think about affirmations in terms of broadcasting energy. In this

way we become transmitters, much like a powerful radio signal. We emit energy through our heart, our DNA, and our brain. Visualize all these components as broadcasting powerful beams of energy, so that anyone who is ready to receive this energy receives it.

The clearer and more focused our energy, the better it can be received. The more true your wish sentence is for you, the more you can identify with it, the more intense your transmitted energy will be.

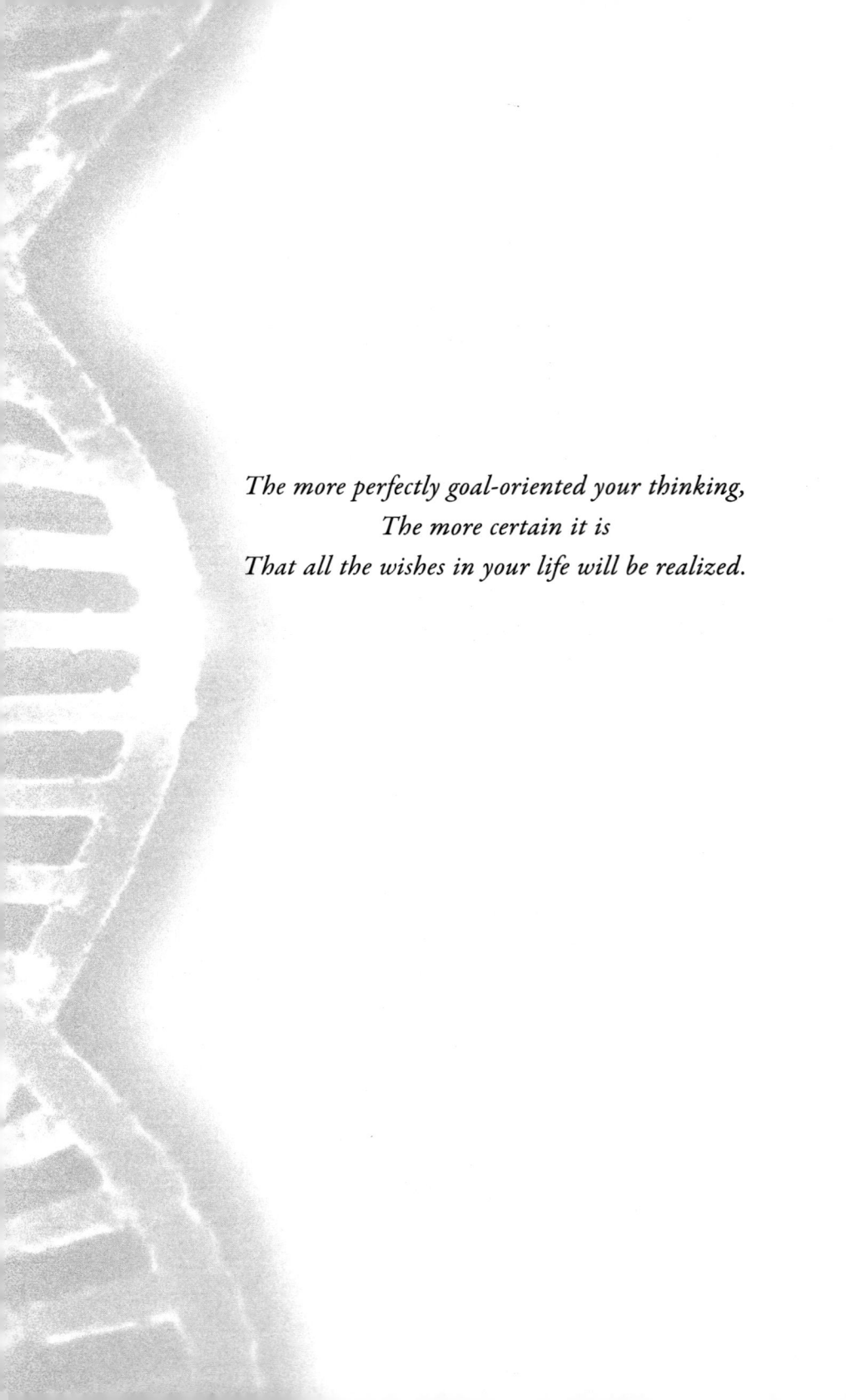

The more perfectly goal-oriented your thinking,
The more certain it is
That all the wishes in your life will be realized.

Build an Image of Your Wishes

Nothing happens unless first a dream.
CARL SANDBURG

Affirmations are a very strong medium that constantly hold us in resonance with our wishes. But there are a lot of other possibilities for sending out the correct energy, the energy we desire, through our thoughts. In general, though, only that which we identify with and which we are occupied with can maintain our generated resonance field.

For some people, wishing is a kind of search engine that trowels for similar energies and draws them in. Since like attracts like, this is how we fulfill our wishes.

The more intensively we occupy ourselves with our wish, the more intense and lasting is the energy that we send to the outside world and, at the same time, send into our subconscious. Maybe this sounds like work, but in reality we should use no effort to do this. Coming into resonance can and should happen *playfully, with ease*. Resonance comes exactly when we are completely relaxed about our wishes. The more easily we can cultivate this state of *relaxation,* the better we will be able to succeed.

There are many ways to achieve this state of total absorption and relaxation around our wishing. We can create a very personal wish image, a type of mental collage of our dreams, a sort of pleasant daydream. Incidentally, this is my favorite way to hold myself in resonance.

Another way is to make an actual physical *successful wish collage*.

- Cut out images from magazines or newspapers of everything that you would like to have in your life, whatever it is. All images, drawings, and photos that are connected to your wish are suitable for this. Don't limit yourself. Include everything, even multiple things. Maybe it is a computer, a bicycle, a house, a dress, a car, an apartment in a certain kind of neighborhood, rollerblades, a purse, a boat, a vacation, a dream partner, cash . . .
- Whatever you would like to have in your life goes into this *successful wish collage*. You can certainly draw or write in there yourself. The choice of format is yours. Essentially, the important thing is that you have this picture—more than a picture, a representation of all your wishes—always in front of you so that you can occupy yourself with it. As you see your desires represented visually, your subconscious will begin to revel in anticipation. You will identify with your wishes. You will be that much closer to your goals, and they will no longer be unattainable. And suddenly they will arrive! You'll find this connection obvious, because what you wanted was already a part of you for a long time. And when it does happen, be sure to welcome its arrival in your life physically.
- Hang this *successful wish collage* somewhere in your house where you step into contact with it daily.
- Maybe something new keeps getting added to your collage, or you draw in some additional details.
- The more your spirit and imagination focuses on this, the more you occupy yourself with it, the quicker it will be drawn into your life.
- The more you anticipate the fulfillment of your wishes, the stronger your transmitted energy will be.
- Connect viscerally with this image. Put a picture of yourself in there, happy and smiling. This is your wonderful future.
- And always remember, fortune is yours!

Naturally, there can be many variations on these *successful wish collages*.

- Draw yourself or paste a photo of yourself in the middle of a page and write honest, positive characteristics around yourself. Also speech bubbles look playful.
- You can note your obvious talents as well as your hidden talents. It is often very impressive and surprising to see how many talents one has and how many positive qualities one possesses. Maybe you can organize well, or maybe you are a good communicator.
- And be sure to write down your goals and visions there! Maybe you wish to become independent, or maybe you want to step into a new career.
- Write, draw, or cut and paste everything into this collage, so that it takes hold of you.

Your successful wish collage is your own personal programming of your resonance field.

Michaela supported and encouraged me to make the kind of collage I describe here, with speech bubbles for myself. Some people call this technique *mind mapping*. I had to write my name in the middle of the picture and around it all of my desired goals and visions. She asked me all the relevant questions to help me find out, from the bottom of my heart, what I *really* wanted. I hung this picture close to the corkboard by my writing desk, so I could see it every day, while I worked or read a book, while I was just thinking, or when I was talking on the phone. Sometimes I was completely unaware what I was looking at, and sometimes I focused on it.

In less than two years, all of my dreams were achieved, even those that seemed to be far off in the future. Because my successful wish collage was located in a convenient spot, every day I was more or less subconsciously connected with my goals, and so something new got organized within me. Some projects even made me fearful. I did not know if I could do them justice. Today they have become a matter of course for me.

It is very effective to make a successful wishing collage for yourself—and it can be very supportive to help others with their collage. When you do so, it is important to let you or your friend's creativity flow freely—just play with it, you will be surprised how many things come to reality! Additionally it is great to make a collage together with your life partner to increase your common visionary power and make your common wishes come true.

Many things are possible in one's life. The less energy—meaning energy as "effort"—we expend, the more relaxed we are, and the more relaxed we are, the sooner we can come into positive resonance.

Create your own personal successful wish collage and see what happens!

One comes into a desired resonance
Only through relaxation.
Only then do wonderful,
Genius solutions arrive.

Accelerate the Construction of Resonance Fields

Start by doing what's necessary, then do what's possible, and then suddenly you are doing the impossible.

ST. FRANCIS OF ASSISI

Remember, when it comes to wishes, our convictions are essential. When we are convinced of something, our heart sends out energy that is 5,000 times stronger than that of our brain and builds a very stable, appropriate energy. And our DNA communicates with everything else that we believe in our deepest heart.

But what happens if we are somehow not completely convinced that our wishes will be fulfilled? We often believe that we are wishing very intensely, very extensively, and in great detail, but we do not notice that along with these visions and aspirations we are also harboring thoughts of limitation, concern that our wishes might go unfulfilled. These kinds of worries are usually hidden in inner questions such as *When will it happen? Am I doing something wrong?*

It's important to note that these kinds of thoughts and feelings are really nothing other than the creative process at work, albeit negatively. We can achieve and manifest our limitations as well as our wishes. We can bring undesired things into our life, we can create a resonance field of deficiency and draw these things into our life, rather than the opposite.

As long as you have mental images and feelings of doubt or impatience with the process, you are occupying yourself with the absence of your wishes and sending those thoughts energy. You limit your wishes with that kind of self-sabotage. You are calling your wishes back before you've had a chance to achieve them.

When we occupy ourselves with deficiency, then we generate a resonance field of lack. When we occupy ourselves with abundance, then we generate a resonance field of wealth.

Sometimes it's difficult for us to be 100 percent convinced of the fulfillment of our wishes without exception. The following, then, is a wonderful method to quickly catapult oneself back into one's own positive convictions. It begins with the usual process:

- Write down your wish on a piece of paper.
- Visually present this wish to your feeling body as an image. How does it feel when this wish is fulfilled?
- When doing this, observe how easy or hard it is for you to create these images. Maybe you see and recognize limitations, how many worries and how much impatience you carry inside, because the wish has not yet been realized.

To come out of this kind of state of self-doubt, it is helpful to give your vision a physical experience. This occurs best with a "convictions game" that I like to do during my seminars. We usually do this in groups of four people. We describe our wish and visualize to the others in the group by sharing that we *already have* what we want. Note, we do not express this as something we don't have and wish we had; we do not say, "I wish to have a wonderful relationship," but rather, "I *have* a wonderful relationship." We tell others how wonderful our relationship is.

As we share in this way with others, our energy manifests more sustainably. Thus we not only convince others of the feasibility of our wish, we are convincing ourselves.

A purely mental experience becomes a physical experience.

This exercise, which always provides a significant acceleration of wishing energy during my seminar, can also be done by yourself.

- Your imaginary powers are strengthened by speaking your wishes in the most beautiful, colorful images, as if you have already attained what it is you wish for. Just do it like professional actors rehearse their part: behave "as if."
- Walk around your house or apartment and speak out loud to the the people in your imagination and tell them how wonderful it is that your wish was fulfilled. Full of persuasive powers, report how wonderful it is, how calming, and how good. In this way you convince yourself. By doing this exercise, you no longer occupy yourself with deficiency or lack. You imagine a life of satisfaction and connect yourself with abundance.
- As you do this, make sure your voice sounds soft, easy, and comfortable. Your body takes in the oscillation of your voice. You begin to identify with the pure physical fulfillment of your wishes.
- If, for example, you want a wonderful relationship, tell these people how wonderful, harmonious, rich, flowing, easy, and cheerful the relationship *already* is. Tell your friends how much you laugh, how sensual and romantic your nights are, how much understanding you receive, etc. Remember, everything you want has already been fulfilled, and you are simply reporting back in detail, in the prettiest, most vibrant colors.
- Similarly, if you want to create a resonance field for a new apartment, report to others how well your apartment is arranged, how nice the neighbors are, how happy you feel in this environment, how exceptional the view is, how much you enjoy the balcony, etc. Provide details.

- Laugh and enjoy yourself as you do this exercise; show your good fortune.
- Finally, sit down calmly one more time and wish for these things again. Go renewed into your visionary powers.
- Now feel the difference. Is it easier for you to create your images?
- Are you now getting more details? Do new colors emerge? Are there scents that you notice? Do new words come to mind? Have your feelings grown? Maybe you notice that the anticipation has become more intense.
- And if so, tell your friends about it.

The more frequently and intensely you do this exercise, the larger and stronger the generated resonance field becomes, and the faster you will realize your desired event.

This exercise is fun, and I have done it frequently, walking around my apartment and sharing with my imagined people all the wishes in my life that have already been realized. In this way, long before these wishes entered my physical reality, I felt anticipation, saw the results with my own eyes, and in this way allowed my convictions to grow and intensify.

Sometimes no matter how much you initially hold this, the mind might tell you that this kind of behavior is childish, but you will be happy with the outcome of this exercise and once you begin to feel and see results, you will enthusiastically participate in this the next time.

To be a creator is to play with energies. The more playfully we deal with our wishing, the more successful we will become.

Using the Healing Power of Sound

Music is a higher revelation than all wisdom and philosophy.

LUDWIG VAN BEETHOVEN

The knowledge of resonance and its effects is ancient. The ancient Egyptians and Greeks knew how sound works. Pythagoras taught that notes and music succumb to rhythmic sequences, and that these oscillations have a special influence on the health of people and animals. He spoke of "spherical music" and developed a harmony that describes the changing effects and connection between our world and the heavenly bodies. Plato said, "Education in music is most sovereign because more than anything else, rhythm and harmony find their way to the innermost soul and take strongest hold upon it."

Given the venerated place that music has held going all the way back to the ancients, it is not surprising that modern science has undertaken many studies on the effects of sound and oscillations on the trinity of body, mind, and spirit.

Hans Jenny (1904–1972), a physician and natural scientist who is considered the father of cymatics, the study of wave phenomena, discovered that every individual cell has its own frequency and its own oscillation. Each frequency is dependent on or in resonance with other frequencies.

The individual frequency or oscillation of each individual cell can be changed through the quality of sound.

One can thus imagine how our entire system reacts to music. When we hear wonderful, uplifting music, every single cell in our body blossoms with a beaming beauty.

The more harmonic and pure the music we listen to, the more we step into resonance with beauty and harmony.

This universal act of harmony is derived from overtones. An overtone is one of the higher tones produced simultaneously with the fundamental tone and that with the fundamental comprises a complex musical tone. Overtones occur in intervals. Any time a harmonic oscillation appears, such an overtone arises. The octave, for example, is the first overtone and is also the most energetic. The quintave is the other overtone. This interval is used more frequently in sacred music. The developing sound stands for universal harmony.

Modern science has found that these harmonic tones help people oscillate in a perfect way, in healthy, healing frequency patterns.

Strictly speaking, the human body is similar to a musical instrument. We therefore possess a harmonic body that tunes itself to the current frequency and changes its form or design accordingly.

- We can quickly harmonize when we hear music that is pleasant to us or listen to musical instruments making this music.
- Singing bowls belong to this group of instruments that rapidly bring about harmony of our mind, body, and spirit.
- When one is not at peace or in harmony with oneself, when you feel overwhelmed, tired, and stressed out, it only takes a few

minutes of hearing the sound of a singing bowl to bring yourself back into balance.
- The wonderful thing about this is that it does not take very long to learn how to use a singing bowl. You lay the bowl in the flat of your hand and take a wooden clapper and glide it along the rim until the bowl begins to oscillate.
- Very quickly you will not only hear one note, you will hear various overtones that also oscillate. Within a few minutes your system will be completely renewed. What for many might seem like a wonder is nothing other than endowing yourself to be in resonance with the holy order and with universal harmony.
- Tuning forks can also be used to create the same effect of relaxation. It has been found that tuning forks with different frequencies can bring the craniosacral fluid, the fluid that is in direct connection with the nervous system, back into circulation. The wonderful thing about a simple tuning fork is that it works on the subconscious level. We do not have to think about it, we do not have to understand anything; the sound itself reorganizes our entire resonance field so that we can again be in harmony with ourselves.
- While listening to the sound of a tuning fork, our thoughts and feelings change, and our entire body shifts into a different stance. Without having to think about it, we can regain harmony within ourselves.
- Using bowls or tuning forks, disharmonies in the body can be neutralized and the body can be brought into resonance with the universal frequency.
- By being brought into harmony, we could suddenly have completely new ideas or find solutions to problems we did not have before. In full relaxation, old patterns and injuries can be contemplated and dissolved.

There is nothing more profound, and nothing can move us as

effectively, as being connected with healing sounds. Maybe you have the desire to get a tuning fork or a singing bowl. Try it when you are tired or stressed out, angry, or out of balance. Then hit a tuning fork, listen to the note, and wait to see what happens. It can be this easy to heighten the vibration in your own resonance field.

Give Yourself the Gift of Recognition

All that we are is the result of what we have thought: it is founded on our thoughts, it is made up of our thoughts. If a man speaks or acts with an evil thought, pain follows him, as the wheel follows the foot of the ox that draws the carriage.

BUDDHA

The higher our resonance field oscillates, the clearer and more positive the energy we send out. The higher the oscillation, the less likely we are to experience troubles, sadness, or hopelessness.

So what is the best and fastest way to raise your oscillation and create an optimal resonance field?

Praise yourself!

This is not always as easy as it seems. We often don't recognize our accomplishments and positive qualities, holding them in hiding. Many of us were taught that it is boastful, arrogant, and immodest to praise oneself, and most of us still carry around this belief. As children, we heard different proverbs to this effect: "Praise in one's own mouth is offensive." "Self-praise is half slander." "One should not praise the day before the evening." "When the mouse is doing too well, it draws in the cat." These and other admonitions intended to shrink our joy, and which served the purpose of bringing us under control, have been the cornerstones of our upbringing.

Eventually we began to assimilate these "rules" as our own truth, thereby limiting ourselves. This behavior is so normal for most of us that we also apply these standards to other people as well. If someone shows unbelievable pride in their performance or if their joy about this is expressed enthusiastically, without limitations, we often judge the person harshly. It is no surprise that we are practically obligated to be self-critical; according to the standards that most of us were raised by, if we did express pride in our accomplishments, we would be perceived as a show-off, arrogant, or pretentious.

So the truth is we are not very comfortable with self-praise.

We have perfected the self-critical view, and we rarely believe others when they praise us. That is why we cannot be proud of ourselves.

What kind of resonance field do you think such behavior and way of thinking creates? Do you think that in this way you could ever draw things into your life of which you can be proud? Probably not. It is therefore time we give up this kind of self-defeating behavior once and for all—at least if we want to create a resonance field that brings positive recognition into our life.

At our seminars I have seen just how prevalent this self-critical attitude is, and how deeply embedded it is in the psyches of most people. When I ask participants to break into groups of four and for an entire minute tell others what is so special and lovable about themselves, the following happens: In the first few seconds, people are shocked and silent. And no wonder—we each have a lot of knowledge and can talk extensively about all the things that are wrong with us, for days, if we have to. But when we are asked to report what is so great about us, it makes us very uncomfortable and we struggle for words.

As soon as the initial shock of being asked to list one's positive qualities wears off, most people begin to laugh. In this case, however, laughing is nothing other than a form of discharging uncomfortable feelings.

After the embarrassed laughter passes, the participants begin this exercise cautiously. Many are very hesitant to start—it isn't very easy to emphasize one's own qualities and share this with others. But it doesn't take very long for this initial tension to be replaced by an unbelievable liveliness in the room. And once the first participant in each group gets over the initial hesitation, it's often the case that she doesn't want to stop self-praising, even after her minute is up. Meanwhile, the energy in the room becomes so high that it is often very difficult for me to get everyone to return to a state of calm.

The participants next switch places, and the next person starts telling the others what is so wonderful about him- or herself. And as we go around to each person in the group like this, the euphoria in the room increases. By the end of this exercise everyone is so playful that they are all laughing and joyful.

So what happened? We each talked for one minute about the things we find good about ourselves. Nothing else occurred. Where does this euphoria come from? Very simply, it is within us. It has always been there, we have only forgotten it. We have kept it a secret from ourselves. We have consciously suppressed it.

When this wonderful energy is released, we feel more powerful than ever before. When freed, this energy spills outward. It truly feels wonderful to be allowed to show others who we really are.

When you want to create a resonance field full of recognition, you can try doing this exercise yourself.

- On a piece of paper, or even better, in a journal, write down everything about yourself that you are proud of.
- Remember, there are many areas in which you are successful.
- Note all your wonderful qualities.
- And then think about it some more. What else can you be proud of if you would allow it?
- Breathe in and out deeply. Read through your list slowly.

- Feel deep within yourself how this feels.
- Feel the power that is within you. Feel your unending pride. Feel how much fun it is to be proud of yourself. It is wonderful to have accomplished so much.
- Add to this list as often as you like. Carry it around with you. And when something else occurs, add it to the list. The more you occupy yourself with these thoughts, the more positive things will occur in the following days.

By doing this exercise you will be astounded by how much of your life has been successful. The more you experience this feeling, the better. These realizations hold your current and future potential. You are capable of being this person today. You *are* this person—you have only forgotten it. You have unconsciously disconnected yourself from this energy. When we separate ourselves from it, it is no longer at our disposal.

Reconnect with this energy. It is very easy. In the moment when you remember who you truly are, in the moment in which you have that wonderful feeling, in the moment when you recall previous images of success, you are reconnecting with your full potential.

- Connect with your past successes.
- You can call this potential up from the past. Do this when you first begin to praise yourself: "I am this. I can do it all. I am capable of it."
- This energy becomes noticeably stronger when you repeat it loudly—best in front of a mirror.

When we do not give ourselves attention, no one else pays attention to us.

There is another step to the exercise I have just described, which further increases the energy. We often do it in our seminars:

This involves telling someone else what you find so wonderful about them. You confirm to your counterpart why you like them, what you find pleasant about them, and the positive qualities you have discovered about them.

This part of the exercise often results in something astounding: after a short amount of time, there are only beaming, smiling faces in the room, and many hugs of gratitude. Everyone in the room suddenly feels close to everyone else. And each person feels closer to him- and herself as well.

Closeness with others and with oneself creates euphoria.

In these moments, as we become closer to one another, we realize we are happy. Within a few minutes, everyone's outlook has completely changed. We need only one single minute to become close to strangers. In one minute, we raise the oscillation of our own resonance field. Although we seemingly directed our outlook toward others, our own resonance field has increased. Why is that?

It is because we cannot see the positive qualities in others that do not also resonate in ourselves. In the same moment in which we want to discover the positive characteristics of other people and focus on them, we also find these characteristics energetically expressing themselves in ourselves and we recall them. When we praise others, we bring ourselves into the same oscillation level of the praise we extend to others.

There is no other way. Because we cannot see in others what we don't already have within ourselves. This, by the way, is true of both positive qualities and negative qualities—we can only recognize in others that which resonates within our own oscillation.

When we recognize the godly potential in others, we see nothing other than ourselves.

- If you would like to increase your own oscillation frequency, focus your view on the wonderful aspects of others.
- If you only look in this direction, your own energy increases abruptly. Always remember: It's all about abundance! Like attracts like. The energy emitted is looking for the same vibrating energy, in this case a field of abundance.
- If you want to realize abundance in your life, go mentally where you recognize your own wealth.

Can One Heal Oneself with the Power of Thought?

You yourself are your own obstacle, rise above yourself.
HAFIZ, SUFI MASTER

Can one be healed through the power of the mind? If you believe the latest scientific research, this question in my opinion can be answered with a "yes." However, we need to approach this subject carefully, for it would be a mistake to awaken false hope, because there are many factors that need to be taken into account to make this true.

Various branches of medicine have become increasingly interested in researching the effects of our thoughts and feelings on our self-healing capabilities. This includes the placebo effect, whereby patients given a placebo—a medically ineffectual treatment—will nevertheless have a perceived or actual improvement in a medical condition. The opposite of this, the nocebo effect,* occurs when someone is given something harmless that actually causes harm. In either case the result, whether beneficial or harmful, is the result of one's thinking and beliefs.

During one of my lectures, a man came to me and told me the following story:

This man had a business with multiple employees. One day, one of his employees was diagnosed with breast cancer. Believing what she'd been told, the woman said, "I will die from this. In one year I'll be

---

*Latin, *placebo*, "I will please"; *nocebo*, "I will harm."

dead." No one could convince her that people have been healed from the type of breast cancer she had. She underwent an operation, which was deemed successful. Nevertheless, she continued to hold to her belief. Exactly a year to the day after she claimed she would die, she died in the hospital.

An impressive example of how strong one's own belief will manifest—in any direction.

The use of one's own thoughts to help bring about healing has been proven to be successful in various forms of alternative medicine. An example is biofeedback, which has been used to cure migraines. These systems use sensors to measure bodily functions, like blood flow in the arteries, or muscle tension. The results are shown on a computer screen. Only when the patient is relaxed do the symbols change on the monitor. The goal is to allow only the right symbol on the screen, which requires truly thoughtful work. Conscious thoughts change the brain waves, and this is made visible on the monitor. In this way, the person learns to influence their own bodily processes through the power of their thoughts. According to a study from the University of Marburg, Germany, this method is as successful as any type of conventional medical treatment.

Through the power of our thoughts, we can be more successful in many areas of healing than we can be with pharmaceuticals.

In recent years there have been many studies undertaken about the mind-body connection and its implications for medicine. According to quantum physics, there are no limits to the possibilities. Everything is possible, even the reversal of difficult, seemingly incurable illnesses.

The following account shows how powerful thoughts can affect the body.*

---

*From Pierre Franckh's *Wunschgeschichten für die Seele* (Wish Stories for the Soul).

Hello, dears,

I was diagnosed with cancer. It has metastasized through my entire body, and there is nothing to be done. The doctors explained that if I am lucky, I will live another six months. I have been told that with chemo and radiation, I can be assured of some "quality of life."

After a few days of introspection, I decided—to the horror of my friends and husband—to accept that my time will pass when it does.

I did not want to die from miserable chemotherapy. I accept my illness and am thankful for the life that has been given to me. I never felt that before. My life was always in the fast lane, work before everything else. Suddenly, this was no longer important.

I set myself up with a program of healing. I moved to the countryside and wanted to use my last time to inwardly "clean up." I began living my life intensely, so that each day was a gift. I no longer lived for my husband or my clients. I stopped working. A friendly doctor supported me mentally. After one year, no metastization or cancer was found.

Kornelia

When one hears about such changes of destiny, it touches us deeply. And it is simultaneously a wonderful encouragement for us to always remember the power of our thoughts.

I found this out for myself when I was just a teenager.

When I was seventeen years old, I had shoulder-length blonde hair. This was my pride and joy, the expression of my youthful revolutionary spirit. It was the time of the Beatles and the Rolling Stones, the time of Woodstock, Jimmy Hendrix, and the Doors. There were the Byrds, the Mamas and Papas, the Beach Boys—and me. I was one of the few in my age group who was allowed to have such long hair. For this, I was envied and admired. I played guitar, had a band, and was immersed in the freedom of the flower power movement.

And then came the catastrophe—my hair began to fall out! I clearly had the beginnings of a completely bald head, which I not only saw as terrible, it also seemed completely ludicrous to me as one of the "pioneers" of the hippie movement in my crowd. I was shocked, heartbroken, and devastated. I secretly searched for solutions in different hair studios around Munich, but no one knew the reason for my sudden hair loss. One person even suggested I try out a costly hair transplant.

In Munich, there was one hair clinic where I had been examined several times. Every time it was found that I clearly had the early signs of becoming bald. The doctors there found this to be completely normal. However, at my age, this was a bit early.

There were nights of tears and despair. I simply could not and would not accept that the symbol of my freedom and youth was lost. But nothing could be done to change it.

And then the following happened:

To earn some money in order to take some psychology courses, which would have been otherwise unaffordable, I helped the organizers of a conference. There I was able to closely observe the presentations of the psychologist Max Lüscher, known for his work with color perceptions, as well as ethologist Konrad Lorenz, author Erich von Däniken, and others. In this way, I got to know a shaman who noticed that day that I had little enthusiasm for the tasks at hand. No wonder—I was depressed. Not once had my mother, who constantly comforted me about my much-discussed tragic event, helped me with these difficult worries.

The shaman asked me what weighed on me so heavily. I showed him the round circle in the middle of my head, surrounded by what remained of my shoulder-length blonde hair. The shaman looked at me with large, surprised eyes. He still could not realize what my problem was.

"Here," I cried with tears in my eyes, pointing to the bald spot on the top of my head.

The shaman only laughed and said, "Why don't you talk to your hair?"

I found the very idea of talking to my hair ridiculous and no longer felt like helping this man prepare his documents.

The next day, he approached me and asked me if I had talked to my hair. I felt indignant at his foolish suggestion, but then I noticed that he wasn't making fun of me, he fully meant what he was saying. He only nodded and said, "You speak with so many people. And I have observed you—you can speak very well with others. Why don't you talk to yourself? Why do you believe the doctors or specialists more than yourself? No one has a greater influence on ourselves than we ourselves. I would never give someone else the responsibility for my own body. I would always want to hold that in my own hands. But that is your decision, of course." He smiled and went back to his work.

That night, I began to talk to my hair, although I found it laughable at first. I told it where it should grow. And I imagined that this bald spot would close up. At least doing this gave me courage.

The next day, the day of the shaman's departure, he saw me and recognized a distinct confidence in my eyes. "Good," he said, "so you have begun."

"Yes," I said, and it suddenly was no longer laughable. "Good," he said again. "But think about it—this is the beginning of a journey. It is only your hair. Do not forget that. It is just dead tissue. Imagine what you could do with living tissue."

From then on, I talked to my hair every night and morning. And suddenly, one week later, I saw the first fuzz appear on my bald spot. Fourteen days later, there was some serious hair growth. In a month's time, I looked funny because the bald spot was overgrown with new hair, but this new hair was still so short that it looked like a short tuft at the top of my head, while shoulder-length hair grew around it.

Half a year later, this haunting episode in my life was all but forgotten.

And today, fifty-five years later, I still have a thick, full head of hair. For me, this is no longer surprising, although I now see what happened in a completely different light. Then I believed that the shaman had helped me a little bit with his magic. Today I know that he only taught me to believe in my own magical powers.

Although I no longer know your name today, dear shaman, thank you for your wonderful help. Who knows how my life would have played out had you not come along?

Although I forgot this incident for quite a while, I have since developed the habit of communicating with my body directly.

When we want to wish ourselves into good health, we don't just wish once and then forget about it; we need to be constantly occupied with this wish.

- If we want to quickly gain health, it is not enough to only wish for health; one needs to also strengthen the entire body through one's thoughts. This is the true catalyst for our wish for health.
- Protect yourself with positive thoughts. Then all self-destroying commands that were previously directed to your body will ease up. All the wrong thought patterns of the past can be dissolved and replaced with new, wonderful ways of thinking. In this way you can quickly find health, as you feed your body with new, fresh, health-furthering energy.
- Imagine exactly how it is when you are healthy again. See yourself with your inner eye as you hop around, ride your bike, play catch, ski, dance, jog, swim, run, or have sex with your partner.
- Mentally paint whatever makes you happy in the most vibrant colors. This strengthens and enormously speeds up your self-healing powers, since you are brought into resonance with a healthy event.

The Power of Attraction between Two Soul Mates

*When there is no agreement about the fundamentals,
it is senseless to forge plans together.*
                                    CONFUCIUS

After health, a happy and satisfying partnership is probably the next most cherished wish. To bring the ideal partner into your life, you need to create the right resonance field to attract that person.

How do two people find each other if continents separate them?
Simple: the attractive power of resonance is responsible for bringing them together. Distance plays no role when it comes to the power of resonance; neither does class difference, cultural difference, or any other hindrances. When two people step into resonance with each other, all seeming hurdles are surmountable. Resonance will pull each person toward the other, oftentimes without their even knowing it, and they are then completely surprised when they find each other and unite as partners.

When two people discover each other, they first recognize themselves.

When we step into resonance with another person, there is an absolute closeness. People do not need words to understand this kind of attraction. Being in resonance with another person is the greatest fulfillment that someone can experience. When this happens, there is the feeling that you have finally arrived—arrived home, to yourself. Because in reality, when this happens you are really just feeling yourself, getting closer to yourself, because the other person is mirroring you back to you. In this case, what we call love is *the recognition of one's own love for oneself,* and at the same time this oneness with yourself creates a sense of oneness with others.

From the beginning, any two people coming together as a couple are searching for similarities. The more similarities they find, the more they believe they belong together, and the more certain they are that their love is everlasting.

We are always looking for ourselves in others.

If we fundamentally hold the same beliefs as our partner, crises or differences of opinion cannot create a crack in our union. Every relationship must endure crises, especially the longer the relationship. But crises can also mean catharsis and healing. That is the deeper meaning of a crisis. We will grow, and we heal unsettled childhood concerns. This can often create a strain and a strong burden on the partnership. But a couple that is in resonance will survive any stress test. Crises are mastered together. Nothing in the world can put to question the feeling of belonging together.

When a couple *not* in complete resonance decides on a path together, this path will be significantly more difficult, since such a union does not possess the absolute attractive power that resonance brings. Both partners will feel that the attractive power is missing. And when it is missing, each person thinks there is a lack of love. In reality, there is a lack of same-oscillating energy.

Why do we not immediately attract our soul mate into our life?
The answer lies within ourselves. Very often, our resonance field is not aligned with that of the soul mate. Sometimes we have completely

different needs. For some people it is more important to be sexually satisfied or to get some form of financial security from a relationship. There may be a fear of abandonment, or one partner may simply want a traveling companion. Maybe there is a fear of closeness and devotion. Maybe there is the feeling that we simply cannot meet a deep, connected love in kind. Maybe there is a worry that if we love again it will end badly. The list goes on and on.

Don't be hard on yourself. We learn through our partnerships, and with every love, we are capable of an even deeper love, until we trust ourselves enough to invite our soul mate into our life.

How Do You Build a Resonance Field for Your Ideal Partner?

Love cannot be given away, when one does not have it; and one has it only when one gives it.
　　　　　　　　　　　　　Aurelius Augustinus

To be in *resonance* means nothing other than being in *unison* with another, to be on the same wavelength as the other. The law of resonance assumes that what happens between two people can only be perceived mutually.

The principle of resonance cannot be one-sided.

When we send our energy, it inevitably meets another energy that oscillates similarly. And when that happens, when we are in resonance with someone else, we feel light, cheerful, exhilarated, euphoric even.

But note that when strong attractive powers are at work, it does not necessarily mean that we are in full resonance with that person, that we have found our soul mate. It can simply mean that a strong similarity has been found. This can mean that we think or feel the same about some known things, that we have similar experiences, or that we share the same sensual desires.

When we are in resonance with another person, we can sometimes feel surprisingly drawn to him or her. The attractive power can be so strong that we are no longer masters of our feelings. They overwhelm us. We cannot escape this. The attractive power can be so intense that we quiver and cry, or that the mere thought of the other person is sexually arousing. Our thoughts are only about this one person.

And then it sometimes happens that perhaps a few weeks or months later, we suddenly no longer feel anything, not a trace of the previous attraction. The magic that so captivated us initially is gone and we wake up to . . . nothing. Maybe we don't understand how we felt so magnetized by this person just a few weeks or months ago. Perhaps we begin doubting or blame ourselves.

In reality, what has most likely happened is that our resonance field has changed. Or the resonance field of the other person has changed, and we no longer oscillate together.

Once our focus realigns to other things, let us say "changed," this vibration or resonance field dissolves and we feel nothing for the perticular person anymore—unless the person makes a similar change and transforms with us.

If the shared resonance field is based only on momentary thoughts and wishes and not on deep, long-term, sustained similarities and beliefs, the partnership cannot last. Thus, in the search for a true life partner, it is essential to generate the *right* resonance field, so that the attractive powers can work to bring us a partnership that can be sustained over a long period of time—a partnership based on a true, deep, loving relationship.

Your resonance field expresses everything to others—constantly and without you being aware.

When you share the same resonance this connection is happening on all levels. Your resonance field communicates not only in a conscious way but also in a subconscious way. This is good, because only on the various levels of both verbal and nonverbal communication can we draw in our ideal partner.

The ideal partner is one whose personality is complementary to our own, because we then have the same goals, wishes, and desires. This means that because we are a certain way, we choose a partner who is mirroring us. We wouldn't want another partner. As well, we will invariably find in that partner both the positive and the negative aspects of our own personality.

If one's pattern is to only seek physicality and not a lot of closeness in other ways, then we will draw in someone who responds to this pattern in kind. We would end up with an affair and not a relationship with depth and trust. At the end of such an affair, this experience—which we have attracted through our resonance—will once again reinforce the feeling that intimacy can only hurt us.

Maybe one of the partners who fears intimacy cracks under the pressure coming from the other partner, who wants something more. This kind of conflict brings the partner who is resistant under so much pressure that he or she either backs out and runs away from the relationship or decides to work through the shadow side to venture into more closeness. Either way, the person must confront his or her fear of intimacy, even if they don't want to deal with it. In such a case, the person's resonance field has sought out the exact person to point out the fear of intimacy. This can be quite painful when it recurs over and over again.

How do you end this kind of painful pattern in relationships?
Certainly not by searching for someone who will "rescue" you. And not by pretending to be someone other than who you really are—that can work only at the beginning of a relationship. Eventually, your fears will surface in the context of the relationship and you will probably become angry that your shadow side has been uncovered and your true character has been revealed.

Likewise, in the beginning of the relationship your partner is also showing his or her best side. But this cannot last very long, and soon enough both partners stand across from one another, disappointed that they have discovered in each other what they've been trying to hide.

The only people who will be attracted to you are those who mirror exactly what you send out.

And so now we understand why we always draw in the same type of person, over and over again. It is only when we change our resonance field that we can be in a position to invite a different type of partner into our life.

How can we invite the right partner into our life?
The best way to avoid all the usual pitfalls is to consider which resonance field we are creating at any given time. One can find this out relatively easily. It only requires that you be truly honest with yourself. Besides, you cannot deceive your own resonance field.

- Examine which people are around you at this time. Then you will know what kind of energy you are sending out.
- Examine your thoughts regarding relationship. Is partnership something that seems wonderful to you or something that makes you feel stressed? Ask yourself, would a relationship enrich me, or would it cause me stress?
- Are you truly open and ready for an intimate relationship, or do you have reservations? Are you really a "relationship person"? Or are you following some sort of program you think you want to fulfill without really knowing whether this is something you truly want?
- Is there anything that you strongly feel you do *not* want to experience? These are precisely the kinds of situations that you will probably attract, because you have not let go of your fears. You are energetically holding yourself in the resonance of your fear. Is it easy for you to imagine being with a partner in a relationship? Or do disturbing images arise when you visualize? These kinds of disturbing thoughts and images are the energies that you are sending out, because they reflect your beliefs. Remember, convic-

tions sent out from the heart have 5,000 times the strength of the thoughts sent out from the brain.

- What are you ready to offer in a relationship? Could you have offered that in previous relationships? Have you allowed yourself to experience this in other relationships? If not, maybe this is something you long for, but you do not truly live this yourself. Instead, you are energetically sending out a deficiency in the very area that you desire.
- Think about what makes you "you." Think about your strengths and weaknesses. Begin to reconcile yourself with your weaknesses, and your future partner will do the same. That means that you have to be aware of your weaknesses.
- You can draw into your life that which you are. As soon as you accept this, you will be an authentic person and will draw into your life a partner that suits who you really are—most of all, someone who isn't hiding something from you, because you aren't hiding from yourself.

It's always best to deal with our shadow side, the side we hide from our conscious awareness, because our resonance field is created not only through our conscious thoughts but also through our subconscious wishes.

How can you figure out what is involved in your shadow side? You can discover this relatively easily. Simply respond with your first thought to the following sentences, without thinking about it. The first response is always correct:

- "I am scared that . . ."
- "I never want to experience . . . again."
- "The bad part about relationships is that . . ."
- "The problem with women/men is that . . ."

Your fears are nothing other than your subconscious wishes, and as we know, fears produce a significantly stronger and more long-lasting attractive power. Everything we don't want to ever repeat is ingrained in our structure, including our beliefs about relationships and our opinions about the opposite sex. All of that shapes the energy we transmit.

Exiting from this negative feedback loop is easier than we think, but to do so we need to observe our shadow side.

Maybe we really want to allow ourselves to experience endlessly deep trust, but we can barely believe that this is possible. If this is the case, then mistrust is the thing that we are bringing into the relationship. And because mistrust is in our energetic field, we will naturally draw in a partner who will inevitably bring this theme to the forefront of our awareness, as long as we have not healed this aspect of ourself.

To a certain extent it is our dark side that determines the basis of any relationship. From there, our wish list—what we really want in a relationship—does not have as much influence in what we attract. More important is the question:

What are you ready to bring into a relationship? What can you give?

Do you want your future partner to always be honest with you, so that your wish list certainly contains "absolute honesty " as one qualification for a partner? Then that is your expectation, and you will not be satisfied with less than that. No cheating, no hiding, no mistrust, no lies, no holding back, no taking things back, no secrets. So far, so good. But you yourself must also bring into the relationship the same qualities you are requesting from your partner.

So then, you must ask the following question of yourself: Can I truly offer that thing I want most in a relationship to the other person? No cheating, no hiding, no mistrust, no lies, no holding back, no taking things back, and no secrets? Do I see myself this way? This is a demand you must make on yourself, otherwise, you are asking a lot from the other person that you yourself are not ready, willing, or able to give. So if you want absolute loyalty and full honesty, you must be able to offer it as well. These qualities should already be evident in you if you want these in another person.

The truth is that we can rarely have everything we want in a relationship. If we want all those things, we must be ready, willing, and able to offer all of them ourselves.

When we don't have something we want, it's because we have not created the right resonance field to attract it.

Something different is oscillating within us. In relationships, this is our true character. The energy of who we *really* are is the strongest sender. It creates our resonance field. For this reason, we only get that which is similar to our true character. Our partner is our mirror, who reflects back to us our own subconscious beliefs.

Do we always get a partner who is just like us?
We always get exactly what we put into a relationship. You may have already noticed this: Here we are in an interesting cycle. On the one hand, we lack most of the qualities we want in a relationship. On the other hand, we need the relationship in order to develop those very qualities, like loyalty, honesty, caring, closeness, trust, sensuality, patience, etc., so that we can experience, learn, and further develop in ourselves that which we really want.

It's a paradox: we do not have the relationship we want because we are missing certain qualities in ourselves. So to develop these qualities we need a relationship.

So how do we get out of this cycle?

This is fundamentally easy. The easiest methods are often the most effective.

- Write down your wishes, and truly write all of them, without any reservations. This list can be endlessly long. Every single detail should make it onto this list. All desires, all hopes, all still-secret ideas.
- The big difference in making this list is the following: do not wish that your future partner brings these things into your relationship. This is not a list of demands for your future partner, it is rather your *development list*. It is your *potential*, your joint goal in a partnership.
- Wish for a partner who is just like yourself, matched to you in all the positive ways but also evenly matched in what you lack. The important thing here is that he or she should also have the same desire to *develop these qualities*.

In this way, your partner will be evenly matched with you. He or she will have the same wishes and desires and the same willingness to work on them. There will be no expectations that you have to compensate for your partner's failings, or vice versa, your partner won't have to compensate for your failings. In this way neither of you has to "audition" for the other. This will relieve all the pressure of expectations. You can be the person you truly are. You can be your authentic self.

In this way you can find yourself in full resonance with your partner. You'll be oscillating together, with the same goals and desires, and with nothing to hide from each other. You'll each know you're not perfect, but you are strong enough to make a path together. Both of you are mature at this point, and open to growing together. Both of you know what it's about and what you want to work on. Neither of you is perfect, and that is absolutely okay because you don't expect perfection in your partner. This kind of forthcomingness in a relationship gives it meaning and a greater purpose.

The fastest way to bring a wonderful partner into your life, a person who is a worthy match for you, who understands you, who can be close to you, a person in whom you can trust, is the person who can meet you on an equal footing.

Wish for a partner who will accompany you in your own growth and development, just as you are prepared to accompany your partner in his or her growth and development.

We can get closer to who we really are when we recognize that we are not perfect, and that we do not have to be brilliant, exceptional, or perfect relationship material, that we can in fact be imperfect and still be worthy of partnership. Only then will we find a partner willing to be close to us. Previously, we did not allow a partner to get close out of the worry that the person would track down all of our secrets and discover the truth of our imperfection. In this way, we kept our partner at a distance.

When we start by being really honest with ourselves and develop a relationship with ourselves based on the truth, we will eventually gain the trust of a partner who is willing to show him- or herself authentically too. Only then do we get closeness, because the game of hide-and-seek is over. And we can finally walk hand-in-hand with a partner, on the same path, and discover the world, because we no longer use up all our energy hiding from ourselves. In this way life becomes very easy. We transform together. We are each evolving into a wonderful soul mate.

Coming into Resonance with Yourself and Your True Feelings

True happiness . . . is not attained through self-gratification, but through fidelity to a worthy purpose.
HELEN KELLER

You will only have those things in your life that you want that are already in you. Nothing else can happen. You will not attract anything else, nor will you even feel drawn to it.

This means that the faster you come into resonance with yourself, the more your world will change in ways that you could only describe as wondrous.

When we are not in resonance with our own true self, our feelings, desires, and wishes are effortlessly carried out by substitute satisfactions. Because we are unaware of our desires or we don't want to fully accept them, we can be easily exploited. The desire for happiness, for example, is such a deep seeded need that it can quickly make a person easy to manipulate. Like a puppet on a string, such a person is readily controlled by various beliefs, people, and institutions.

In fact, there is a gigantic industry that operates on the principle of exploiting the desires of people for profit. The advertising industry distracts us from our true, deep desires and shows us that we will quickly find happiness if we succumb to any of myriad temptations. The consumer boom, with its false portrayals of happiness, controls us once our sense of happiness is tied to these temptations, which try to convince us that we can only be happy if we consume.

But as long as we stay in resonance with our true feelings, it is easy for us to recognize which things and people will truly enrich our life. We will no longer fall for these deceptions, since we recognize them for what they are. Our consciousness will completely filter out the fake bids on our perception. We will know that all of these things surround us, but we will no longer feel drawn to them.

So how do we come into resonance with ourselves? Check out your answers to the following questions:

- What do you need to be happy?
- Okay, now what do you need to be *truly* happy?
- What can you not live without?
- What do you like about yourself?
- What can no one know about you? What "must" you hide?
- What do you hide about yourself?
- What are you ashamed of?
- What do you disapprove of?
- Which qualities are those you do not dare express fully?

All of your answers to these questions reveal your true desires and fears. The more you reject them, the more you distance yourself from yourself. The more you stand by them and claim them, the more authentic you are.

Stand by who you really are, including all facets of your being.

We each have a dark side. But who sees this as a shadow? Maybe our light is also found there, because our creativity, healing, or our survival hides behind this.

You are wonderful. You are unique.

Yes, you have taken on burdens on your journey. Love yourself for your burdens. Love yourself for your detours. This is what makes you "you." It's what makes you a human being—a wonderful and unique human being. If you were perfect, then you would already be enlightened and no longer here on this earth plane. And truthfully, what is interesting about a perfect person? Perfect people only make others feel inferior.

The more you stand by your dark side, the more you accept it, the more readily you'll be able to bring it into the light and the happier you'll be.

When you can stay by yourself as you really are, you arrive at yourself.

Welcome to the club of the "unfinished"! When we no longer have to "audition" or pretend, all the sham battles come to an end. We can allow other people to be just as they are and be easy on ourselves. There is no greater gift than that of being who you truly are. Stand by all aspects of your being.

Using the Power of Prejudgment in a Positive Way

The man with a new idea is a crank until the idea succeeds.

MARK TWAIN

We often ask ourselves if we can consciously influence others through our resonance field. Can we actually cause others to change their behavior through our thoughts?

An interesting study was shown on television recently. Researchers were trying to determine how strongly future performance was influenced by common prejudices.

To find an answer, the researchers sought to find out how blondes would fare on this test. Would their results be worse than those of women with a different hair color? Would the results validate the stereotype that blondes are simpleminded or dumb? Or would they be able to hold their own and do as well if not better?

(Before you read any further, try to answer this question for yourself.)

The result of the examination is certainly as striking as enlightening: If the person is given the test without seeking to influence them beforehand, the blonde average is on a par with the others taking the test. But if this test opens with a tangential question, such as: "Are you a blonde or brunette?" and there is a corresponding box to check off, then memories of all the prejudices that have been heard and endured over a lifetime are evoked, and this actually affects their current behavior.

In this case, blondes that are confronted with blonde stereotypes at the beginning of the test did significantly worse than brunettes, also worse than the blondes who were allowed to take this test without this "destabilizing" beginning.

The effects of these prejudices were even more serious for people who took a comparable test in the U.S.: Black participants had to check off whether they were white or black at the start of the test. These test participants actually performed significantly worse than other black participants who were not asked this at the beginning of their test.

Simply reminding the individual about a prejudice causes him or her to immediately fall into this trap, even when the individual seeks to ignore, deny, or combat it. Everything the individual has heard for decades on a routine basis is triggered at that very moment. In other words, we immediately resonate with these prejudices and submit to them.

Consciously or subconsciously, we are always drawn toward what others say about us.

The same study showed that when blondes are encouraged about all the things they do well, they perform even better at these skills than previously thought possible. Therefore,

- If we want to do well on tests and successfully complete examinations, it is important that we not only tell ourselves how good we are and that we can achieve anything we want, it is also important to be in an environment that reinforces this idea.
- Seek mental support from your friends and family.

Prejudices do not always have to be verbalized or shown clearly. It is enough that we *think* about someone in a certain way to affect a change in behavior.

I recently read about a test given by a behavioral research institute, in which a boy was given mathematical problems to solve. Before the first test, the supervising math teacher was told that the testing boy was a mathematical genius. For this reason, the teacher was sympathetic to the boy for the duration of the test. As a result, the boy was in a very good mood, and his brain, which had been connected to electrodes that measured his brain function, worked extremely well.

For the second test, the math teacher was changed, but the boy was the same boy as in the first test. This time, the math teacher was told that the student was learning disabled. As soon as the boy walked through the door and got ready to sit down for his test, the electrodes measured a blockage in his brain. Only through the unspoken prejudice that the math teacher harbored against the boy did this measurable blockage present in the boy's brain. The electrodes clearly showed that the brain was working at a reduced capacity. And, of course, the boy did not do nearly as well on the second test.

Prejudices are nothing other than strong, purposeful beliefs that have the power to influence others through the transmission of energy.

We can hinder or encourage other people through our thoughts.

The power of our opinions on others can be shown through a test I give frequently in my seminars. This test stems from applied kinesiology and is frequently used by mental trainers of a Bundesliege soccer team to show the players just how much their own opinions affect their teammates.

In this test, I ask one person to come to the front of the room and hold their arm parallel to the floor. First I test the basic force. During this, the person is asked to pull their arm upward with force as I try to push their arm down. Depending on how strong the person is on this day, this succeeds well or not. If the person feels weak in some way that day, I then ask them to concentrate on a positive affirmation—like "I can do it"—to be sure that this person is in her or his "full power" and show to the audience the effect of this.

Next, I stand behind the person so they cannot see me, and I show the seminar participants what they should think about this person. If I give them a thumbs-up, all participants in the lecture hall should only think the best and most wonderful things about this person. If I give them a thumbs-down, they should only think the worst about the person.

What happens next surprises the seminar participants every time I do this. When I give the thumbs-up while standing behind the person, and all the attendees think wonderful things about this person, the person's arm stays powerfully up and through my best efforts I cannot push it down. If I give a thumbs-down, I only need two fingers to push the arm down. All the strength disappears from the arm.

If we think positively about someone, we strengthen them. If we think negatively about someone, we weaken them.

This knowledge is meanwhile also used in high-performance sports.

The mental trainer of the Bundesliege soccer team showed in a television show how strongly the opinions of the players affect one another's moods. If the teammate believes that another player will successfully take a pass, or if they are convinced that a dribble will be successful or that they are in the position to shoot a goal, then this affects their mood. Only through the beliefs of others does a player gain confidence and power and also the ability to perform the task.

Positive prejudice or prejudgment can act encouragingly, such as when we believe our partner or we give our child our full confidence. But the following is also true: pretending is useless. The resonance field cannot be deceived. Only when we truly believe it and are convinced of the skills of others do we give power to others.

- If you give your child your confidence in her power, she too will believe in her own skills.
- If you are full of trust in the ability of your partner, his trust in his own abilities increases.
- If we try to fake it, the exact opposite will occur.

It is in our hands to change our environment. We influence it through our own opinions and beliefs.

- Consider whom you give your trust to.
- Whom do you influence with your thoughts?
- Whom do you hinder through contempt or mistrust?
- Look at your environment. How do you participate in creating what's around you? Does it maybe correspond to your unconscious beliefs or expectations?

We control our environment with our beliefs much more than with our thoughts. So if you would like to change your partner, then change your opinions and beliefs about him or her. If you want harmony in your family, then believe that this harmony can exist and see it as a potential. Emit this kind of loving energy and you will see that your fellows also have a choice to react in a patterned way or respond creatively. When you change your opinions about other people, they have a chance to show you the best sides of themselves. More importantly, this gives them an opportunity to change their opinion of you. Always remember, like attracts like. Resonance fields oscillating together complement each other. When you keep this in mind, you will get more strength and trust from others.

The more we give, the more we get.

So why not use the power of prejudice or prejudgment *positively*?

Can One Bring About Enduring Change in the World through Positive Thinking?

If we did all the things we are capable of, we would literally astound ourselves.

THOMAS EDISON

We can improve the resonance field of others through our own oscillations. But can we go one step further? Through our beliefs, can we influence not only the qualities of individuals but also those of an entire group of people?

The answer is a resounding *yes!*

It has been extensively documented just how powerful our thoughts are and how little is required for them to change the environment. The following are just two examples.

In 1974, Maharishi Mahesh Yogi, the founder of Transcendental Meditation (TM), undertook a study conducted in twenty-four American cities with populations greater than 10,000. He explained that violence and break-ins significantly decreased when only 1 percent of the population in a specific area practiced TM. This meditation practice is well-known for producing the experience of peace. Maharishi assumed that the inner peace experienced in the participants of this meditation would be mirrored in their environment.

The results of this experiment were more than impressive. Although only 1 percent of the population in any given area participated in the study, the change was clearly measurable. As long as the meditation continued, offenses significantly decreased, by an average of 16 percent. There were fewer break-ins, fewer robberies, less violence, and even fewer murders. The emergency personnel of hospitals had less to do.

And this was not only in *one* city, these results were evident in all eleven cities. Although these results could not be scientifically explained, they could not be denied. This phenomenon of rising coherence in the collective consciousness has since been described as the Maharishi Effect.

A few years later, a similar project under the direction of the late Dr. Charles N. Alexander was conducted in July and August 1983. In this experiment, the conditions were much more difficult. Tensions were extremely high in the Middle East, and there was a brutal civil war going on in Lebanon. The experiment created a group of resident Israeli TM meditators in Jerusalem to test the effects on crime and the quality of life in Jerusalem and Israel as a whole, and a group in Lebanon to assess the results on the war in Lebanon.

Eight social indicators were used in this experiment: (1) crime in Jerusalem, (2) crime in Israel as a whole, (3) automobile accidents involving personal injury in Jerusalem, (4) fires in Jerusalem, (5) a stock index of all freely traded stocks on the Tel Aviv stock exchange, (6) a national-mood scale derived from content analysis of a major newspaper, (7) reported war deaths of all factions in the Lebanese war, and (8) a war-intensity scale of the Lebanese war derived by newspaper content analysis.

As in the 1974 experiment, the group of meditators selected in Jerusalem, Israel as a whole, and Lebanon were calculated as representing approximately 1 percent of the total population of those areas. And again, as with the U.S. study, the results were impressive: crime decreased by 7.4 percent in Jerusalem and by 4.1 percent in Israel overall, while the quality-of-life indicators rose. The war in Lebanon was also significantly impacted, as seen by a significant decrease in war deaths and war intensity during the period of the experiment. When the period of meditation ended, the statistics rose to their former levels. The result of decreased war due to groups practicing TM has since been replicated in seven other similar experiments.

The truly surprising thing that these experiments revealed is that fewer people are needed to increase peace in an area than one would think: approximately 1 percent of the population or less. That means that with a population of one million, only 100 people are required to shift the consciousness of the entire group, and our entire world population of approximately seven billion people would need only about 85,000 meditators to shift the consciousness of the entire planet!

When only a small portion of the population creates inner peace within themselves, this is quantitatively mirrored in the environment.

This is explainable in terms of quantum physics and bioenergetics and supported by the results of the aforementioned experiments and others like these: that the power of our own beliefs potentiates when a group has common beliefs and radiates these outwardly.

Our beliefs can change the world.

Our beliefs already do this constantly, right now, in this very moment. But *what* do we believe? How many of us believe that we cannot change this great big world, with all its complexities, problems, and people? "What can *I* do to affect seven billion?"

We naturally think this way at first. It even seems obvious. But we also need to know that this is only one belief. And beliefs have the power to shape the world. When we follow the same beliefs as others, that all of the problems of this world are not solvable, we move downward very quickly, and the violence of this world only increases, because our thoughts are working collectively to further our negative expectations.

- Do not take the beliefs of others and simply accept them as your own, because these beliefs are furthered with your acceptance. Think for yourself!
- If you are convinced that you matter in this world, you will soon meet other people who believe the same thing.
- Think about it: it takes only 100 people in order to change a group of one million!
- It's important to note that those 100 people don't need to be physically in one location together—they could be spread out all over the world.

For a long time now my website, www.pierrefranckh.com, has been a forum where people can interact with one another. One of the participants came up with an idea to start a "wish day" with all of the participants, so that at a given point in time everyone would think of their wishes and send out positive energy—an example of a powerful group of people with the same beliefs, who wish for and think positive thoughts together, at the same time. Every Friday at 8 p.m., all the participants meet mentally wherever they are and mentally connect with one another and put their collective wishing power to work. At the end of these sessions they explain what they accomplished and what happened to them. The feedback has been touching, astounding, and wonderful. This section is, incidentally, the most-read on my site. Developing a common consciousness happens much faster than one thinks.

You are important. You can influence this world. But only you can decide what that influence will be.

Like footprints in the sand,
The footprints of a long-gone life
Remain within us.

Although those who created them
Leave new footprints elsewhere
Their prints are still within us.

We ourselves leave such prints
Every day
Within ourselves
Without consciously perceiving it.
Yet they still determine our thoughts, feelings,
And our beliefs.

Instead of clearing the footprints out of our souls,
We relive them
And our unconscious wishes
Over and over
In our dreams and desires.

Set on our stage,
We let our characters,
Our scenarios, play out,
Scenarios that fit our worldview
Even when they are not our own,
Even when we don't like them.

But they are only footprints of a time past
That we do not want to let go of.

How Can Old Beliefs Be Transformed?

However difficult yesterday may have been, you can always start anew today.

BUDDHA

When we do not attain our wishes, there is usually a subconscious belief that is stronger than the wish. This belief then works against the wish and is often longer-lasting, with significantly more strength than the wish. This belief is often disguised in the form of a worry or another firm belief. Most of these disruptive convictions do not stem from us at all. They are usually the convictions of our parents, grandparents, siblings, teachers, friends, or acquaintances. Not infrequently they may also be the convictions of our pastors, schools, television programs, or advertisements. These convictions often concern moral ideas, ideas about what is right and wrong, what we should think is beautiful and ugly, what is good and bad, and what we should accept and reject.

Strictly speaking, every person who has had any type of decided role in our life has a part in shaping our beliefs. For a long, long time, ever since we came into this world, we have been subjected to others' views about who we are and how we should behave in our world. We learned very early on to judge and condemn. That which we condemn today is mostly based on the concept of morality as passed down by our parents, acquaintances, teachers, and friends. We still try to live out their beliefs and visions. But we not only evaluate others, we evaluate ourselves, too. And most of the time, we evaluate ourselves the same way our parents saw us.

And so we come to all these beliefs, such as "I will never get it right," "I never have enough money," "I am a loser," "I am not pretty enough," "That doesn't work on me," "I cannot do anything," "I do not believe that I will become anything," "Others are much better, smarter, faster," "That's just the way I am and I'll never change." If you think like this, try asking yourself who offered you these kinds of proclamations at an early age, until they finally became your reality.

- Write down all these old scripts and subconsciously repressed beliefs to bring these patterns into the light.
- Next, write down sentences that you heard a lot as a child, and also those that bring back negative memories—all the terrible, humiliating, and hurtful statements that you had to hear as a child, like "You can't do that"; "Let me do that, you can't"; "You numbskull"; "The way you look, you'll never find a man."
- Give yourself time. Everyone has these kinds of statements in their past. Simply remember and note them.

Your list could be very long. Don't worry, you're not alone. Most of us have comprehensive lists of negative pronouncements that were handed down to us from childhood. Although it may have been many years ago that you were bombarded by these kinds of statements—thirty, forty, or even fifty years—it is still astounding how these declarations reverberate through our life today, even though we've matured.

The process of writing down these statements often brings to the forefront buried memories, things we thought were long past us. Upon closer examination, we will recognize that very often these judgments are still quietly at work within us. And although we may never have been fully convinced that the thoughtless statements of our parents and teachers were true at the time we first heard them, we nevertheless picked up these affirmations of our supposed inadequacy just by hearing them, and most likely we've carried them forward subconsciously.

I too heard many of these kinds of utterances as a child and took them as truth. Even the very first statement my mother made to me as she held me as a newborn stuck with me for decades. I was barely in this world when I was greeted—she much later shamefully confessed—with the following pronouncement: "You are not really the prettiest child."

This statement became my truth for many years and made for powerful feelings of inferiority. My mother certainly did not say this in a mean way, but it nevertheless was a formative opinion. It became my truth. When others later found me attractive, sexy, or good-looking, I did not believe them. When anyone encouraged this opinion that I was truly more a type of "character" and that I was not particularly handsome, I could unconditionally agree with them.

So you are not alone in having heard all the negative judgments made about by various authority figures when you were a child, and which then turned into your reality.

It could be that while writing your list you become sad or angry, or that you feel tired and your mind wants to wander away from this painful subject. Allow all your feelings to come to you and stick with this task. You have been carrying these feelings around for a long, long time.

- After you have written down all the negative statements that have defined your life, begin to formulate them into something new and positive.
- You can change sentences such as "You can't do that" into "I can do anything I want."
- Or "You'll never find a man" can change to "I am a wonderful partner in a loving relationship."
- "You only bring bad luck" will change to "I am a gift for everyone in my life!"
- "You're fat" turns into "I love myself the way I am."

When we change these negative beliefs into something positive, something very deep happens. Our understanding begins to form anew. We learn that there is an alternative to what we previously understood as being truth.

When we ask ourselves who said that to us then, we often determine that the person who said these things probably never meant it personally but was caught up in their own insecurities and the beliefs that were imposed on them when they were young. Maybe your mother was caught up in her own problems. She might have been overtired, overworked, or impatient, or had just had a fight with your father. Maybe your father had financial difficulties or was simply overwhelmed by his life situation. Very often, adults simply repeat the same statements they heard as children, and in so doing they carry on the tradition of uttering these kinds of negative statements to their children—us.

The good news is that we can break this generational cycle now. We can accept that when we were children certain people did not recognize our creativity, our curiosity, our liveliness, our beauty, and our potential because they were being governed by their own scripts, which had been passed down to them from the generation before them . . . and on and on. As soon as we open our eyes to this, it is much easier to heal from the negative consequences of these judgments. This is where it is useful to formulate positive statements, or affirmations.

We now have a chance to release and transform all the old, negative patterns, just let go. This is much easier to do when we don't harbor any resentment toward the person who said these things to us; instead, we return these statements to their authors full of love and affection for them.

- Picture in your mind's eye surrendering the list of negative statements made by these people—what they said to you—and then turning around and simply walking away.
- We should think that the people who once said these things to us did the best they could. They were not capable of more at

the time. That does not mean that they are good or bad people. They are simply people, with all the mistakes and limitations that human beings have. They just didn't know any better then.

As long as we carry around resentment and anger toward those who harmed us through their negative patterning, we cannot rid ourselves of the negative beliefs they imposed on us.

It is therefore best to focus on positive statements.

- Feel the strength and joy that comes from these affirmations. Identify with them and you will notice how the effects of negative thoughts and pattern sets have influenced you.
- The most effective method here is a little ritual in which you burn the old pattern sets that you wrote down. Remember to do this in a safe place.
- When you do this, feel deep down within yourself that you have just released old, invalid thoughts and that they have left your life forever. Let the feelings of lightness and liberation that result from releasing these old scripts move up and through you.
- Fill the places left empty by these departed beliefs only with your positive affirmations. Concentrate on them. Speak them loudly and clearly. Let these become your new convictions. The more you feel their power, the faster the new resonance field is created.

We can change our life, sometimes in just this one simple act. However, it is essential that we give some time and room for these new experiences to grow within us. We should repeat these newly formulated beliefs as long as needed, until they are anchored deeply in our conscious mind. The changes will sink in faster than we may think possible.

One of our seminar participants shared with me that her life changed completely on the evening of the day in which our group burned the old pattern statements together and replaced them with new, positive affirmations. Her sample statement was that she was not lovable and was not worthy of anyone being sweet or nice to her. As a result, she found herself in a marriage in which her husband always mirrored those exact thoughts back to her and withheld tenderness from her.

Sometime after that burning ritual, she related to us that without sharing with her husband the ritual she had done at the seminar, her husband, completely unasked, took her in his arms one evening and tenderly caressed her and held her while she cried. And she awoke still in his arms the following morning.

When we change our resonance field, our entire environment changes. We often do not need any words to accomplish this. It is more effective to change our own resonance field, our oscillation. With that, we do not risk losing our courage to face someone, yet we can still affect their behavior. And then suddenly, situations change in wondrous ways. In reality, we know that it is no wonder—it is our own preparedness and courage to stand behind our ability to change and transform.

Every Forgiveness Brings a New Beginning

One is not only responsible for what one does, but also for what one does not do.

LAO-TZU

Many say that letting go is not very easy. True—but it is possible. But what often prevents us from letting go of the old is a single point: the "I cannot forgive" point.

If we cannot or will not forgive—particularly those who have deeply hurt us—we can never be at peace with the past. It's often the case that we want those who have done these things to us to pay for it. The bad news is, this probably never will happen. We simply bear the consequences ourselves.

We impede ourselves from truly living life until we have forgiven.

This is neither good nor bad. No one can force us to forgive someone else. And yet if we don't, we pay the price.

If, for example, we often think about injuries or injustices we have suffered and do nothing to release this preoccupation, we will experience similar injuries and injustices over and over again because we are bringing this energy into our life through the law of resonance. If we cannot forgive and in so doing release our attachment to this oscillation, *we* continue to carry the true effects of these injuries, even if it doesn't seem fair that the very party that has been wronged—

us—should continue to feel the pain of the original injury. We may think that the person who inflicted so much pain should atone for what they've done, should at least pay for it, but most likely the person whom we cannot forgive may be long gone, or perhaps he forgot what he did to us, began a new life, and lives happily now.

When you cannot let go and you hang on to sadness and revenge, you are the only one who is paying the price. You become trapped in the resonance field of the original hurt, and your environment will align itself to it. The world will seem unfair to you. And experiencing more unfairness—the result of being trapped in that resonance field—only strengthens the belief in how unfair the world is. Maybe all you hear are very sad stories that further your gloomy view that not only your life but the lives of others are like that too. As a result you gravitate toward like-minded, similarly resonating people, because . . . Well, as we know, like attracts like.

Ultimately, it is always your own decision as to which resonance field, which reality, you want to create.

We all know people who for years, sometimes decades, have not forgiven something. We can read this inability to forgive on the physiognomy of the person. Years-long embitterment buries itself in the face of the person. The person often rewinds the same tape, caught in the same thought loop that no one wants to hear anymore. We all know these people are always lonely and isolated. And very often their thoughts are accompanied by chronic illness.

Let's say somebody did something bad to you. Maybe it was unfair and mean, deeply hurtful or malicious. Do you want to give this person even more power over you in the present and into the future? When we cannot forgive, we have made a decision to continue the hurt that was done to us, and by focusing on that hurt, we give the person who hurt us our power. Maybe the person has been dead for a long time, or they have since created a happy life and have long since forgotten having hurt us. In such a case, only we are still caught up in the cycle of irreconcilability.

When you do not forgive others, they still have power over you.

As long as you do not forgive and let go of old injuries, you will have no new, wonderful experiences in your life. You will remain in the resonance field of revenge and sadness, and you will draw these kinds of experiences into your life.

Therefore, it is essential that we allow ourselves, after a period of mourning, to realign again in our life and move forward to build new visions. Don't give the past more power, otherwise it will determine your life. Of course we are allowed to be sad, because only the sadness helps us release the relevant emotions.

Sometimes old injuries return after decades and we're not sure why. Even when we have forgotten the original cause of the hurt, we still hold on to the emotions of hurt. These emotions can be slippery and difficult to grasp. It can be hard to recollect the original trauma you want to forgive and release, which is why it can be helpful to do this forgiveness exercise:

- Write down all the injuries you have ever experienced in your life to the best of your recollection. Do not be surprised if you release a lot of pent-up energy through this exercise. It could be that you become angry or sad all over again. If so, let it happen and do not judge your feelings.
- Write down every single detail you can remember. Maybe it was unimportant to others, but it is essential to you.
- Do not limit yourself. Everything that arises has legitimacy.
- Give yourself some time to do this. Maybe you need multiple days

for this. When you begin to think about this, you will begin to open long-closed channels of your memory.
- Be aware that every single person alive has been subjected to injuries. You are not alone.
- Review this list when you are at peace. Many of the things you have written about probably happened a long time ago. Maybe you forgot many of these things, and yet you have not let go of them and started your life anew.
- Now ask yourself: Which one of the people on your list do you want to still have power over your life? And who do you think you can let go of?
- Place a check mark next to the people and experiences that you want to part with. Parting, in this case, means to be free. With each forgiveness, one finds a new beginning.
- Decide which themes you want to hold on to. Be open and truthful. It makes no sense to try to fool yourself—your resonance field cannot be betrayed.
- Remove from the list whom or what you can forgive. Simply cut it out with scissors.
- Think about what you learned from each person or experience you have cut off your list. Be thankful that you had these experiences and people in your life to teach you something valuable, even though it may have come in a painful package.
- Now, take all these paper strips and place them in a fire-safe bowl and burn them in a safe place. Remove these people and experiences from your life without resentment, with gratitude, because you have become richer and more conscious through them. Say good-bye, with love and affection, to these "companions of destiny" and wish them a wonderful life.
- Feel how freeing it is, how you can now breathe more fully, how this release has opened new spaces within you.
- Hold on to this new consciousness during the whole day until you go to bed.

Do not be surprised if one or more of these persons from your past contacts you. Others will energetically feel when they have been released from someone's life. Take this as a sign of how effective your work has been. The faster you can let go, the faster you will be able to change your resonance field, and heretofore unknown benefits and offers will come to you as a result. Sink into them, even if they are uncomfortable and new. You finally have the possibility of going forward and choosing from an unlimited number of new paths.

Change Your Past

He who wishes to read the future must leaf through the past.

ANDRÉ MALRAUX

Many claim that the past cannot be changed, but the future can.

To a large extent, the future results from our past. So it would be very good if we could at least affect our past a little bit, right? Quantum physicists say that we can do this. Furthermore, they have effectively proven it.

We can change our past and build a new, more comfortable future for ourselves. In fact, we do this constantly. But unfortunately we often change our past to our disadvantage and therefore influence our future negatively. How is this possible?

As we know, there is no absolute truth. There is only a subjective truth. Each of us will take in the same experience differently and then store it in our memory. Napoleon Bonaparte said, "History is a set of lies that most people have agreed upon." Similarly, our memory is a subjective truth of the past, on which we have come into agreement with ourselves, and it may not correspond to the actual truth. Oftentimes the scenes transpire in our imagination only as we subjectively perceived them. Maybe we were hurt at specific times in the past, or felt humiliated or rejected; we then judge everything else according to this feeling. Those who keep a journal are familiar with this, because the memory of an experience changes when it is written down. Sometimes people make it prettier, or they simply see it differently than the way it truly happened.

The memory also changes when we retell an experience over and over again. It gets a different coloration from day to day. It becomes prettier and more colorful when we report it to others. We add sentences that we supposedly said, weave in smart arguments or actions, and are always the hero of our story. Or we paint the things that negatively impacted us with even more dramatic color.

We unconsciously also change our past, our personal past, daily, because with time we start to believe our own words and our own version of the story. Today we identify with and resonate with this changed form of our memories. We then accept these as truth and allow this truth to become our reality. This quite human behavior forms an essential backdrop for our future.

American theoretical physicist John Wheeler, a colleague of Albert Einstein and a proponent of Einstein's unified field theory, determined that the mere observation of things has modifying effects on our behavior. He came to the conclusion that even pure observation is imaginative. And no wonder. We already have preconceived notions, feelings, and convictions about everything we observe. We evaluate. And these judgments are enough to generate resonance fields through which we connect with our world.

When we consider our life, our behavior, our well-being, or our need for existence, our career, our big achievements and failures, our partnerships, then we are seeing our convictions, aligned with everything in our past, mirrored back to us.

Through reconsidering our past, we can change our future.

We can naturally use our creative processes consciously, in that we can replay the past anew. Or said differently, this is how we can transform our thinking about the past.

For example, we can consciously reflect back on the benefits of past situations, even negative ones: What have we learned from them? How strong have we emerged from them? Where did this person or event take us? Could we lead the life we lead today if things had gone completely differently? Would we possess the same power or endurance that we have today had things gone differently?

You have become the wonderful person you are today through the crises and traumas that you have lived through. This is true for everyone. It is precisely these past traumas and losses that have opened your eyes and have taken you to a place where you have become more courageous, where you are capable of taking brave new steps into the future. After all, there is no life without disappointments. A life without problems or difficulties is only available in our imagination.

Life presents us with many hurdles. But hurdles are not negative, not even if we tend to think of them that way. Hurdles help us become stronger and more powerful. Only when many problems and crises are resolved in your life can you be relatively confident and calm when presented with new challenges, which you can then solve with greater ease.

Past crises afford a great potential for us.

We must truly contemplate so that we do not remain stuck in old injuries, but rather we use the potential that has since developed in us as a result of those hurts. Only then can we change our own story of our past into a small success story. And this story is not a lie; it speaks the truth. We *have* achieved it. We *are* wonderful, extraordinary, superb, and we have moved forward and are open to what the future brings. When we consider our life in such a way, we realize that we have developed our life further with each step on our path.

And then suddenly we find we have this great, wide-open potential at our disposal. We resonate with the successes of our life. This means that we send our potential for success into the universe and draw further success into our life through the law of resonance.

Our future is changed by the way we look at our past.

- Recognize which skills you have developed in the past.
- Note the strengths you demonstrated while experiencing crises.
- Which new paths and skills have resulted from those crises?
- What would you not have experienced if you wouldn't have had disappointments in your life?
- Write down everything that made you stronger.
- Write your own success story. Write down the wonderful story of your life, with you as the hero of the story. You have done so many impossible things, and you have always managed to get back up and show your true stature. You have surmounted hurdles and coped with problems.
- Be proud of yourself.
- Add to your success story as often as you feel like. Occupy yourself with it.
- The more time you spend working on your success story, the greater your resonance field of success. You are that successful person!

Our personal history can change, mainly in the way we retrospect. Our future is created out of our observation of the past. What use we get out of this recognition lies entirely in our own powers of choice.

- Decide now, in this very moment, to create a completely new resonance field.
- This power is within you, and it has always been within you. Maybe it has been lying dormant and has only waited to be awakened by you.

Do Not Overstrain Yourself

What lies behind us and what lies before us are tiny matters compared to what lies within us. And when we bring what is within out into the world, miracles happen.

HENRY S. HASKINS

Begin your new life, with its new resonance, carefully and with loving patience, so that you don't completely overwhelm yourself. Better to go into your planned changes one step at a time.

Begin with the projects that come the easiest, because with them you will gain the courage for bigger plans. Never let your goals out of sight, but also do not feel bound to address them on the spot. When you keep your future projects alive in your thoughts, you do not have to force anything; instead, in a playful way, new possibilities will come to life, doors will open, and helpful people will appear at your side. When we step into resonance with our wishes, we also make decisions and weigh new steps, but it becomes significantly easier for us to do so. If you think you have to immediately change everything, you'll likely become overwhelmed because you haven't given yourself time to grow into your vision and your transformed resonance field cannot be consciously maintained because it isn't completely stable yet.

We can naturally achieve anything we want to achieve and draw it into our life, but the desired goal is not always the auspicious one for us, so beware of what you wish for. Sometimes our own desire does not do us justice; the feeling of inferiority or the feeling of not being loved prevents the actual happiness. However we go on little by little, growing with our task. We develop our own modifications in which we often reach the results faster than we previously thought possible. Above all, we are happy. We are happy along the way to reaching the destination. We find, as we have reached our goal, that new projects will come to us and we discover new, exciting, and larger visions for ourselves going forward.

We are constantly in motion and will always be that way. A well-known proverb says the only constant in life is change. But we can influence and direct the changes in our life. It's like driving a car: we find enjoyment in the ride, not only in arriving at our destination. Otherwise, we suffer for the entire ride.

If we enjoy ourselves on our way, and if we feel secure at the helm of life, we can steer our life in any direction we wish.

- To be sure what you want to change or develop in your life, it is helpful to write down all the things or skills that you want to develop.
- Think big! Do not limit yourself, regardless of whether it's "difficult" or "impossible" to have these desired changes in your life.
- Write down everything and be as detailed and precise as possible. It is best if you can also visualize these images.
- What is the "bigger picture" in your life?
- Think in terms of cycles: Where do you want to be in five years? Ten years? In twenty? Imagine the best vision for yourself that you can.
- Resonance fields that we maintain over a long period of time develop great strength. Think of the long-term effects!

*If you're in resonance with your true desires,
you need not fight or expend a lot of force.
You are carried with lightness and happiness.
You are secure in your own abundance.
That's what is for you.*

Our deepest fear is not that we are inadequate.
Our deepest fear is that we are powerful beyond measure.
It is our light, not our darkness, that most frightens us.
We ask ourselves, Who am I to be brilliant,
gorgeous, talented, and fabulous?
Actually, who are you not to be?
You are a child of God.
Your playing small does not serve the world.
There is nothing enlightened about shrinking
so that other people will not feel insecure around you.
We are all meant to shine, as children do.
We were born to make manifest the glory of
God that is within us.
It is not just in some of us; it is in everyone
and as we let our own light shine,
we unconsciously give others permission to do the same.
As we are liberated from our own fear,
our presence automatically liberates others.

MARIANNE WILLIAMSON

AFTERWORD
Simply, Thank You

Dear friends, I thank you for the wonderful trust that you offer in sharing your stories in e-mails and letters. I hope to be able to continue to do justice to them. These are true gifts—and you give me plenty of them and I sincerely thank you for that!

I thank my wonderful wife Michaela Merten, my soul mate, best friend, and my inspiration in life.

I thank our beautiful daughter Julia Franckh, who taught me what unconditional love is and who provides us with the latest news on cognitive science and many more interesting findings.

Thank you to all my wonderful supporters, beloved ones, and our families.

This book is dedicated to all great visionaries who constantly live their dream.

Resources

More information about me and my activities can be found at my website, www.pierrefranckh.com. Have fun!

Pierre Franckh and Michaela Merten's Weekend Seminars

I will be giving weekend seminars in the United States. Please check my website for schedule information. The following are topics I help participants address in my seminars.

- How do I come into resonance with my wishes?
- How do I wish correctly?
- How do I create a positive resonance field the fastest?
- How do I give my wishes power?
- How do I recognize my unconscious resonance field?
- What attacks my conscious wishes and how can I change it?
- How do I let go of my worries?
- How do I track down all my thought patterns?
- How do I clear the way to allow for my wishes to manifest?
- How do I further develop my wishes?
- How can I shape my life to make it wonderful for me?
- How do I achieve happiness in my life?
- How do I develop my goals in my work and in my partnerships?

By addressing personal questions and concerns during the seminars and hearing what other people have to say, this gives deeper insight into one's own behavior around existing wishes and shows possibilities for how one can step out of the cycle of one's own negative patterns to achieve a better quality of life.

Once we sense the power of our wishes and our personal strength to change things in our life according to our will, we not only get back our feeling of self-worth, we also feel much more balanced. When we begin to successfully implement our wishes and goals, then we feel happy. We feel like an active participant in the world that we shape according to our desires. We go from powerless dependence on others into independence and freedom to create the world we want to live in.

Creating a positive resonance field changes our entire world: our experiences, our partnerships, our approach to life, and our manifest reality. When we not only understand the principle of successful wishing but actually experience *that* it works and *how* it works, it can change our entire life.

Wonders happen every day. Why not also to you?

Coach Training with Pierre Franckh and Michaela Merten

The coach training with Pierre Franckh and Michaela Merten is aimed at all who want to work as coaches in our method. This training is intended to be integrated in your existing consulting services.

Coaching is exciting and challenging work. You can support people in their professional and personal development and be part of their changes.

In coach training, you too will also change and develop.

Only when a person goes through the coaching process and has further developed him- or herself can they successfully coach others. With this comprehensive education, you will get the tools you need to completely support other people professionally.

Each training consists of five intensive seminars, held over the period of one year. For more information, go to:

www.pierrefranckh.com

or contact us at:

Wolfgang Gillessen
Schönstr. 72b
81543 München
Phone: +49 (0) 89-680-70702
Email: **wgillessen@t-online.de**

Index

abundance, 109, 110
affairs, 133
affirmations, 14, 95–96
   how they work, 41–43
   power of, 98–101
   using, 98–101
Alexander, Charles N., 149
Anka (case example), 37–39
ankylosing spondylitis, 26–28
anticipation, 83
Army, U.S., 30
athletes, 77, 78
atoms, 53–54
attention, 119–20
attraction, 9, 128–30
Augustinus, Aurelius, 131
Aurelius, Marcus, 56

Backster, Cleve, 31
bad luck, 90
bank closures, 87
Beethoven, Ludwig Van, 112
belief systems, 2–3
   affirmations and, 98–101
   interaction with physical world, 13–14
   outside world and, 17–21
   transformation of, 153–58

biofeedback machines, 77–78
biographies, 80
Bonaparte, Napoleon, 164
Braden, Gregg, 31
brain
   connection with heart, 12–14
   malleability of, 93–96
   mirror neurons and, 75–80
Buddha, 41, 52, 68, 79–80, 116, 153

case examples
   Anka, 37–39
   Kornelia, 124
   Maria, 59–60
   Monika, 42–43
   Sabine, 20–21, 33–34
   Sandra, 25–26
catharsis, 129
cells, 23–28
climate change, 87
collages, 104–6
communication, 35–39
Confucius, 128
conjugate complex waves, 46
connectedness, 17–18, 56–60
connection, 6
consciousness, 50

convictions, 12–13, 55, 56–57, 95–96
Cramer, John G., 46–47
craniosacral fluid, 114
creative visualization, 47–48, 79
creativity, 58, 71
crises, 129, 166
criticism, 117

Däniken, Erich von, 125
distance, 6
distant DNA, 30–34
Divine Matrix, The, 31
DNA, 6
   communication with environment, 35–39
   distant DNA and thought, 30–34
   impact on physical world, 18–21
   influence on feelings, 23–28
doubt, 65, 109
dualism, 17–18

ease, 64, 103
echo waves, 46, 49
Egyptians, 112
Einstein, Albert, 10, 30, 36, 45, 93
electromagnetic energy, 12–13
e-mails, 91
emotions, 12–13
empathy, 77
empowering affirmations, 100–101
emptiness, 57
energy crises, 87
energy fields, 6
entertainment, 89–90
epigenetics, 2
era, breakthroughs of, 7

euphoria, 120
event probability, 46, 50
exercises
   attention, 119–20
   collage, 104–5
   godly potential, 120–21
   making the mental physical, 110–11
   self-praise, 118–19

faith, 14–15, 54, 99
famine, 90
fears, 57, 84, 87–88, 135–36
feelings, 140–42
   connection with heart, 10–11
   DNA and, 23–28
forgiveness, 161–63
Francis, St., 108
Franckh, Pierre, 87–91
   self-healing and, 26–28, 124–27
frequencies, 112–13
friends, 71–74
fun, 65–66
future, 47–49, 164–67

Gandhi, Mahatma, 79–80
Gariaev, Peter, 18, 35
goals, 105, 133
Goethe, Johann Wolfgang von, 75, 81
gratitude, 25, 120
Greeks, 112

Hafiz, 122
harmony, 113
Haskins, Henry, 168–69
healing, 124–27

heart
    as center of emotions, 10
    energy field of, 11–12, *11*
    wishing and, 15–16
heart coherence, 23
HeartMath, 10–12, 13, 23
history, 164
hopes, 57
human belief, affirmations and, 41–43
Huxley, Aldous, 23, 35
hyperspace, 36

imagination, 93
inferiority, 70–71
influence, 68
Institute of HeartMath, 10–12, 13, 23
intimacy, 133
Israel, 149

Jenny, Hans, 112–13
Jerusalem, 149
Jesus, 14–15, 80
journals, 91
joy, 25, 64, 65–66

Kafka, Franz, 82
Keller, Helen, 140
King, Martin Luther, Jr., 80
knowledge, 81
Kornelia (case example), 124

lack, 109
Lao-Tzu, 159
laughter, 117–18
law of resonance, 6, 8–9, 22, 33, 58
Lebanon, 149–50

letters, 91
like attracts like, 9
loneliness, 57–58
Lorenz, Konrad, 125
love, 25–26, 29
Lüscher, Max, 125

Maharishi Mahesh Yogi, 148–49
Making of Reality, The, 46
Malraux, André, 164
Mandela, Nelson, 80
Maria (case example), 59–60
mathematics, 2
matter, 9, 52–54
Max Planck Institute, 95
McCraty, Rollin, 23, 30–31
media, 87–91
memory, 88–89
mental training, 78
mind mapping, 105
miracles, 70–71
mirroring, 133, 134
mirror neurons, 75–80
Monika (case example), 42–43
motivation, 71–74, 82–83
movies, 89–90
music, 90, 112–15

Nägerk, Valentin, 96
negative feelings, 24–25
negative resonance, 71–73, 108–9
negative thoughts, 58
neural correlate of consciousness (NCC), 96
neurons, 94–95
news media, 87–91

nocebo effect, 122
nonlinearity, 47–49
normal quantum waves, 45–47

obstacles, 80, 82
octaves, 113
offer waves, 46, 49–50
Olypmic Games, 77–78
overtones, 113

patterns, ending, 133
Pavlov Institute of Psychology, 32
personality, 133
perspectives, 58–59
pharmaceuticals, 123
photons, 18, 53
placebo effect, 122
placenta, 24
Planck, Max, 44
plasticity, 93–96
Plata, Manitas de, 76
Plato, 112
playfulness, 103
Poponin, Vladimir, 18, 35
positive resonance fields, 74
positive thoughts, 26, 127, 148–52. *See also* thought
possibilities, 2–3
potential, 73–74
poverty, 90
praise, gift of, 116–21
prejudgment, 143–47
present, 47–48
prestige, 79
pride, 119
prophets, 14

propositional waves, 46
prosperity, 82
Pythagoras, 112

quantum field, 19–21
quantum physics, 2
quantum waves, 45–47
quarks, 53

reality, 7, 35–39, 47–48, 94
recognition, 25
  gift of, 116–21
Rein, Glen, 23, 30–31
relationships. *See* soul mates
relaxation, 103–4, 107
resonance, law of, 6, 8–9, 33, 58
resonance fields, 6, 65–66
  accelerating construction of, 108–11
  as hinderance to development, 68–74
  news media and, 87–91
  for soul mates, 131–39
  surrender and, 82–85
  true feelings and, 140–42
respect, 25
Richo, David, 85
Rizzolatti, Giacomo, 75
Rosen, Nathan, 36
Russian Academy of Sciences, 17–18

Sabine (case example), 20–21, 33–34
sadness, 65
Sandburg, Carl, 103
Sandra (case example), 25–26
self, 140–42

self-doubt, 70–71, 109
self-healing, 124–27
seriousness, 64
shadow side, 135–36
shamans and shamanism, 125–27
singing bowls, 114
sleep, 88–89
Socrates, 48
soul mates
    attraction between, 128–30
    building field for ideal partner, 131–39
sound, healing power of, 112–15
spherical music, 112
Stark, Johannes, 53
Starkmuth, Jörg, 46
stress, 88–89
success, 82–83, 119
surrender, 82–85
synchroncity, 85

tension, 64
Teresa, Mother, 80
terrorist attacks, 87
Thompson, Jeffrey, 32
Thoreau, Henry David, 17
thought
    cells and, 23–28
    creating new future, 45–50
    healing and, 122–27
    impact on distant DNA, 30–34

matter and, 52–54
world change and, 148–52
time, 6, 45, 47–49
Transcendental Meditation (TM), 148
tuning forks, 114
Twain, Mark, 143–47

unemployment, 87

vibration, 8
visions, 105
visualization, 79

Waitley, Denis, 77–78
wealth, 79, 121
Wheeler, John, 165
will, 56
Williamson, Marianne, 171
winning, 78–80
wishing concept, 7, 10–16, 32, 61
    affirmations and, 98–101
    becoming reality, 35–39
    building an image, 103–6
    resonance field and, 64–66
wishing energy, 52
Wolf, Fred Alan, 49
world change, 148–52
worm holes, 36
worry, 84

Zeeman, Pieter, 53

Books of Related Interest

The Science of Getting Rich
Attracting Financial Success through Creative Thought
by Wallace D. Wattles

The Council of Light
Divine Transmissions for Manifesting the Deepest Desires of the Soul
by Danielle Rama Hoffman

Discover Your Soul Template
14 Steps for Awakening Integrated Intelligence
by Marcus T. Anthony, Ph.D.

Energy Medicine Technologies
Ozone Healing, Microcrystals, Frequency Therapy, and the Future of Health
Edited by Finley Eversole, Ph.D.

The Basic Code of the Universe
The Science of the Invisible in Physics, Medicine, and Spirituality
by Massimo Citro, M.D.
Foreword by Ervin Laszlo

The Heart-Mind Matrix
How the Heart Can Teach the Mind New Ways to Think
by Joseph Chilton Pearce

Molecular Consciousness
Why the Universe Is Aware of Our Presence
by Françoise Tibika

Morphic Resonance
The Nature of Formative Causation
by Rupert Sheldrake

INNER TRADITIONS • BEAR & COMPANY
P.O. Box 388
Rochester, VT 05767
1-800-246-8648
www.InnerTraditions.com

Or contact your local bookseller